An Illustrated Atlas of the Skeletal Muscles

Bradley S. Bowden
Alfred University

Joan Bowden
Alfred University

Medical Advisory Board

Bradley Boyd, D. C.
Karen Kalbach, B. S., L.M.T.
Mark B. Reid, M.D.

Morton Publishing Company
925 W. Kenyon Avenue, Unit 12
Englewood, CO 80110
800-348-3777
www.morton-pub.com

Book Team

Publisher	Douglas Morton
Project Coordinator	Dona Mendoza
Design/Production	Ash Street Typecrafters, Inc.
Cover Design	Bob Schram, Bookends

Printed in the United States of America

10 9 8 7 6 5 4 3 2 1

ISBN: 0-89582-616-X

Preface

The **Illustrated Atlas of the Skeletal Muscles** is designed as a functional reference and study guide for students and health professionals. Some of the most unique features in the Atlas are the "information windows" that include brief descriptions of representative problems resulting from injury and disease. Palpation and trigger points are also included in the "information windows" in the chapters that illustrate the muscles. Trigger points are signified by a bullet (●).

The first chapter presents the human skeleton, with bones and bone groups clearly labeled. The chapter will also identify "bony features" that will be used in subsequent chapters to define the specific points of attachment for muscles and the locations of bones in specific types of "joints" or articulations. Bone and muscle features are labeled by numbers that correspond to the numbers in the key on each page. This enables the user to hide the key and use the diagrams as a quiz. The chapter concludes with illustrations of the primary bony palpation sites.

Also featured in Chapter One is an 8 page full-color insert illustrating the human systems. It includes the skeletal, muscular, circulatory, respiratory, digestive and urinary, and endocrine systems.

In Chapter Two, the Atlas illustrates the primary types of articulations between bones in different parts of the body. This enables the user to thoroughly understand the degree and range of motion resulting from muscle contraction.

Chapter Three illustrates the diversity of movements in different parts of the body. Inclusion of the articulating skeletal elements in the body reinforces the names and locations of bones, the association between articulations, the specific sites of muscle attachment, and muscle movements.

Each of the muscles chapters are presented in color with their specific points of skeletal attachment clearly indicated. The origin, insertion, action, and nerve innervation of each muscle are listed along with palpation and trigger points and other practical information.

The Atlas also includes illustrations of the muscles of the eye and the tympanic cavity along with illustrations of the tongue, larynx, pharynx, and palate.

Finally, Chapter Eleven presents a series of illustrations of nerve innervation pathways to major muscles and muscle groups. This enables the student or professional to appreciate the potential relationship between muscle weakness and damage to specific nerves.

Acknowledgments

We want to thank Douglas Morton and Dona Mendoza who first proposed this project and convinced us of the need for a skeleto-muscular atlas of this type. As project manager, Dona Mendoza kept everyone on task and on target, attending to the innumerable details and coordinating work between the authors, the illustrators (Peggy Firth, Chris Creek, Scott Annis, and Christopher Allen) and Joanne Saliger, the compositor. We would like to extend a special thank you to Peggy Firth, the medical illustrator who created the majority of the illustrations in this atlas.

We particularly appreciate the valuable comments, suggestions, and criticisms of the members of the team of medical professionals assembled by Morton Publishing Company who reviewed the manuscript.

The Authors

Contents

7 Muscles of the Shoulder and Upper Arm   137

8 Muscles of the Forearm and Hand   159

9 Muscles of the Hip and Thigh   191

10 Muscles of the Lower Leg and Foot   217

11 Muscle Innervation Pathways   243

Glossary   259

Index   262

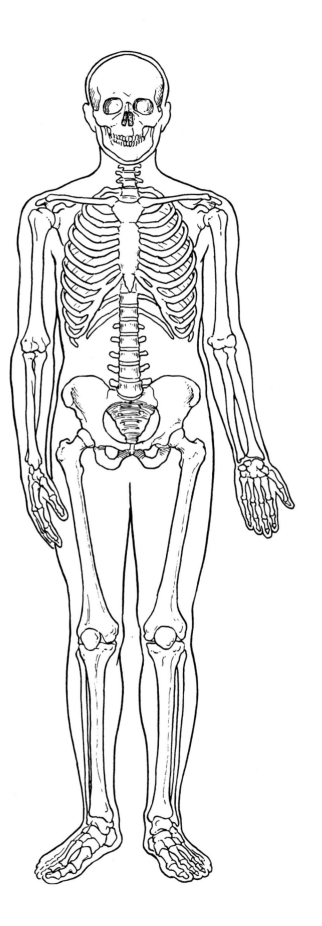

1

The Skeleton

Anterior and Posterior Views of the Skeleton

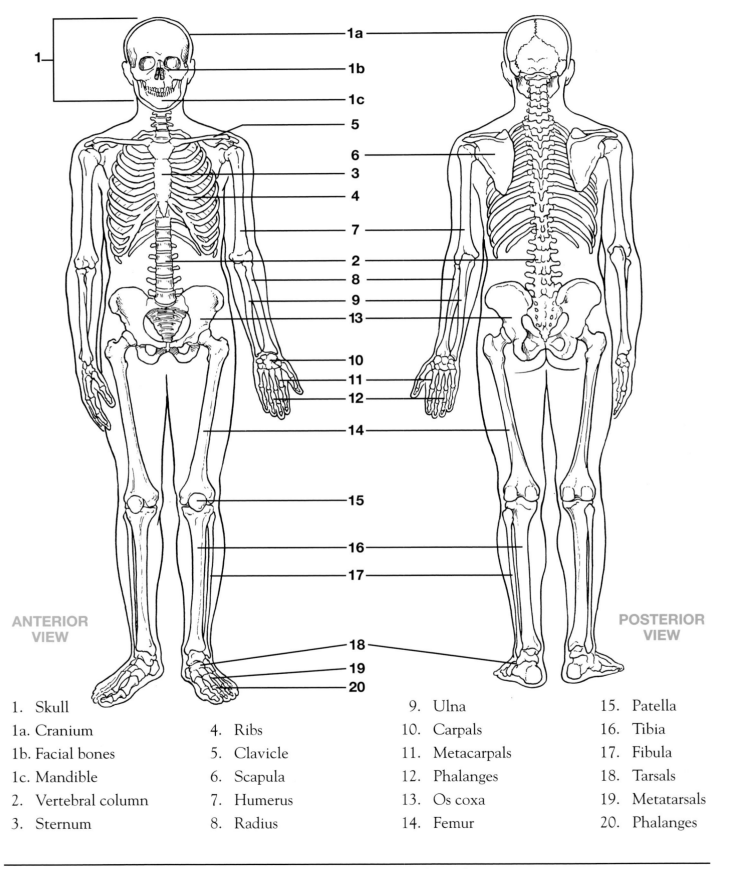

ANTERIOR VIEW

POSTERIOR VIEW

1. Skull
1a. Cranium
1b. Facial bones
1c. Mandible
2. Vertebral column
3. Sternum

4. Ribs
5. Clavicle
6. Scapula
7. Humerus
8. Radius

9. Ulna
10. Carpals
11. Metacarpals
12. Phalanges
13. Os coxa
14. Femur

15. Patella
16. Tibia
17. Fibula
18. Tarsals
19. Metatarsals
20. Phalanges

Lateral and Anterior Views of Skull

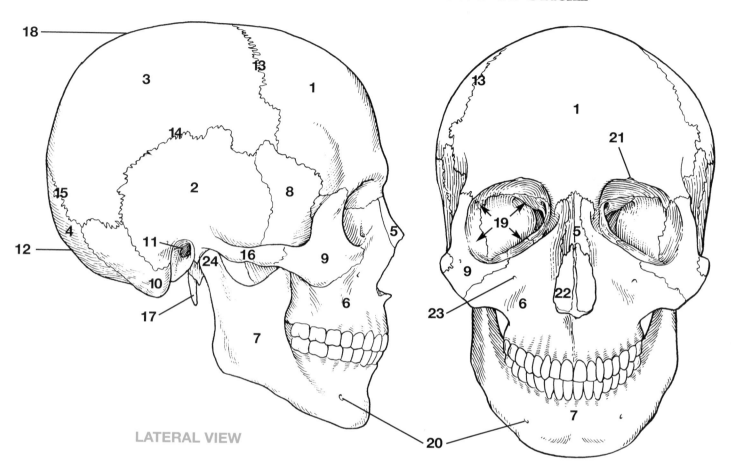

LATERAL VIEW

ANTERIOR VIEW

1. Frontal bone
2. Temporal bone
3. Parietal bone
4. Occipital bone
5. Nasal bone
6. Maxilla
7. Mandible
8. Sphenoid bone
9. Zygomatic bone

10. Mastoid process of temporal bone
11. External auditory meatus
12. External occipital protuberance
13. Coronal suture
14. Squamosal suture
15. Lambdoidal suture
16. Zygomatic process of temporal

17. Styloid process of temporal
18. Sagittal suture
19. Bony orbit
20. Mental foramen
21. Supraorbital foramen
22. Nasal septum
23. Infraorbital foramen
24. Mandibular condyle

The **temporomandibular joint** usually dislocates anteriorly. During yawning or taking a large bite, contraction of the **lateral pterygoid** muscles may cause the heads of the mandible to dislocate. In this position, a person is unable to close his mouth. Dislocation can also occur during tooth extraction. Most commonly, the dislocation is caused by a blow to the chin when the mouth is open.

Superior and Inferior Views of Skull

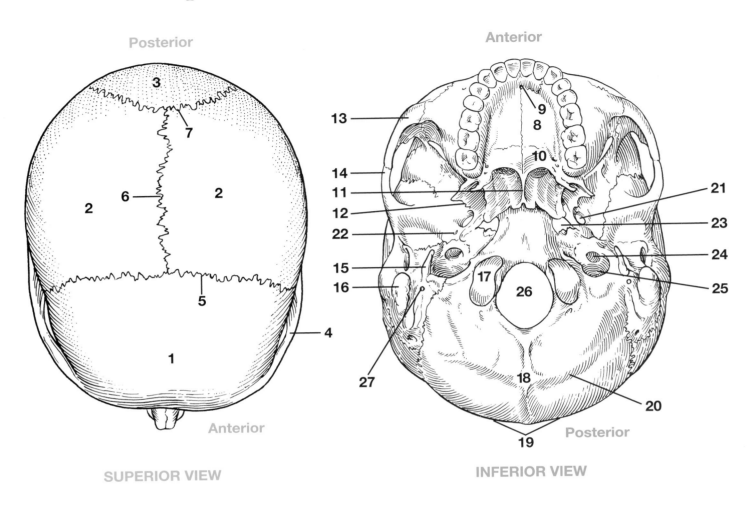

Posterior

Anterior

SUPERIOR VIEW

INFERIOR VIEW

1. Frontal bone
2. Parietal bone
3. Occipital bone
4. Zygomatic arch
5. Coronal suture
6. Sagittal suture
7. Lambdoidal suture
8. Palatine process of maxilla
9. Incisive foramen

10. Palatine bone
11. Vomer bone
12. Sphenoid bone
13. Zygomatic bone
14. Zygomatic process of temporal
15. Styloid process of temporal
16. Mastoid process of temporal
17. Occipital condyle
18. Occipital protuberance

19. Superior nuchal line
20. Inferior nuchal line
21. Foramen ovale
22. Foramen spinosum
23. Foramen lacerum
24. Carotid canal
25. Jugular foramen
26. Foramen magnum
27. Stylomastoid foramen

Internal View of Base of Skull

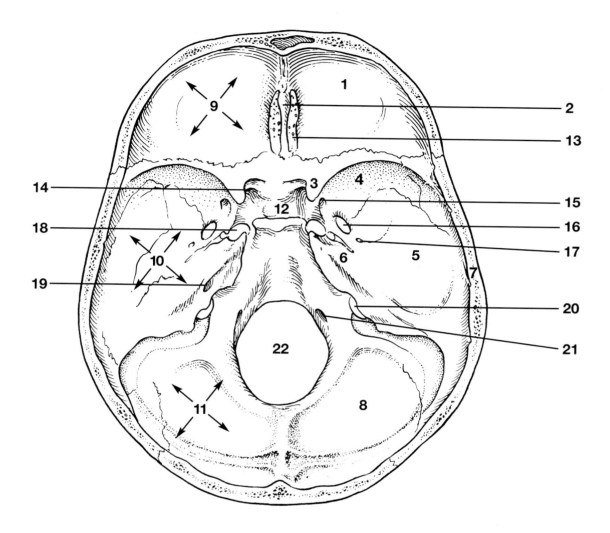

1. Frontal bone
2. Crista galli of ethmoid bone
3. Lesser wing of sphenoid bone
4. Greater wing of sphenoid bone
5. Temporal bone (squamous part)
6. Temporal bone (petrous part)
7. Parietal bone
8. Occipital bone
9. Anterior cranial fossa
10. Middle cranial fossa
11. Posterior cranial fossa

12. Sella turcica of sphenoid bone
13. Cribiform plate of ethmoid bone
14. Optic foramen
15. Foramen rotundum
16. Foramen ovale
17. Foramen spinosum
18. Foramen lacerum
19. Internal acoustic meatus
20. Jugular foramen
21. Hypoglossal canal
22. Foramen magnum

Cervical Vertebrae, Hyoid Bone, and Thyroid Cartilage

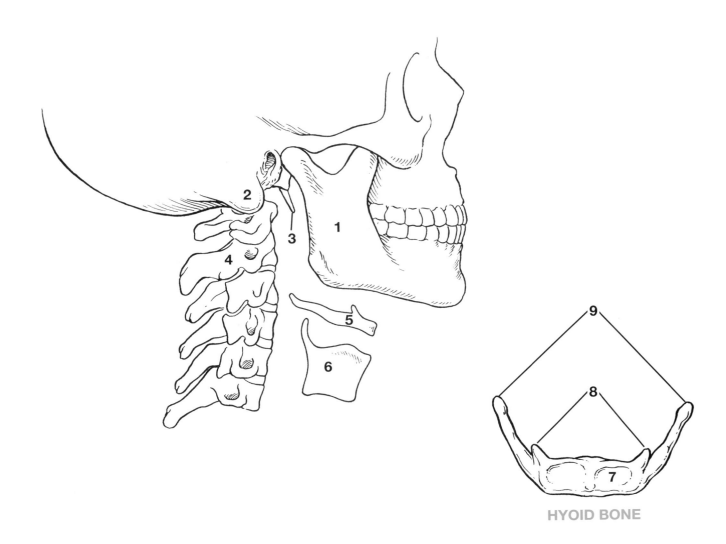

HYOID BONE

1. Mandible
2. Mastoid process
3. Styloid process

4. Cervical vertebrae
5. Hyoid bone
6. Thyroid cartilage

7. Body of hyoid
8. Lesser cornu
9. Greater cornu

The **hyoid bone** does not articulate with any other bone. It is suspended by ligaments from the styloid processes of the temporal bones and serves as a site of attachment for tongue and throat muscles. The hyoid bone is often fractured in incidents of strangulation and is, therefore, carefully examined during an autopsy in which strangulation is suspected. The **thyroid cartilage** is the largest of the nine laryngeal cartilages. The two main plates of this cartilage are fused in front to form a laryngeal prominence (Adam's Apple) which is more pronounced in males and females after puberty. This cartilage can become fractured as a result of blows received during boxing, karate or compression by a shoulder strap during a vehicle accident. The protective guards hanging from ice hockey goalie masks offer protection against this type of injury.

Sternum and Thoracic Cage

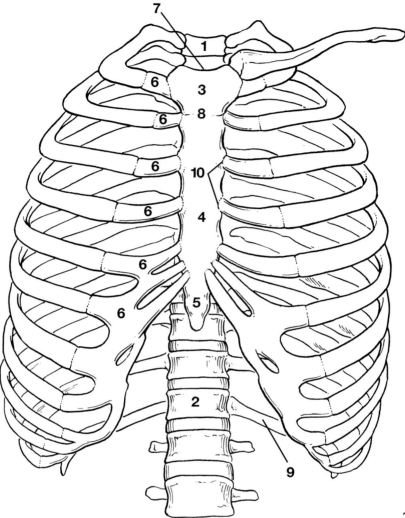

1. First thoracic vertebra
2. Twelfth thoracic vertebra
3. Manubrium of sternum

4. Body of sternum
5. Xiphoid process
6. Costal cartilage

7. Jugular notch
8. Sternal angle
9. Floating rib
10. Costal notches

Three important anatomical landmarks of the **sternum** are the jugular notch, the sternal angle and the xiphisternal angle. The **jugular notch** is the midline depression at the top of the sternum. It is approximately in line with the point from which the left common carotid artery branches from the aorta. The **sternal angle** can be felt as a horizontal ridge across the sternum. It is in line with the intervertebral disk between the fourth and fifth thoracic vertebrae. It is often used as a reference point for locating the second rib and for listening for heart valve sounds. The **xiphisternal joint** marks the beginning of the xiphoid process. When performing CPR on a patient, it is important not to press on the xiphoid process which might break and puncture the underlying liver or heart. During the procedure called **sternal puncture,** a needle is inserted through the sternal surface into the red marrow to aspirate a sample of red bone marrow for laboratory analysis.

Vertebral Column

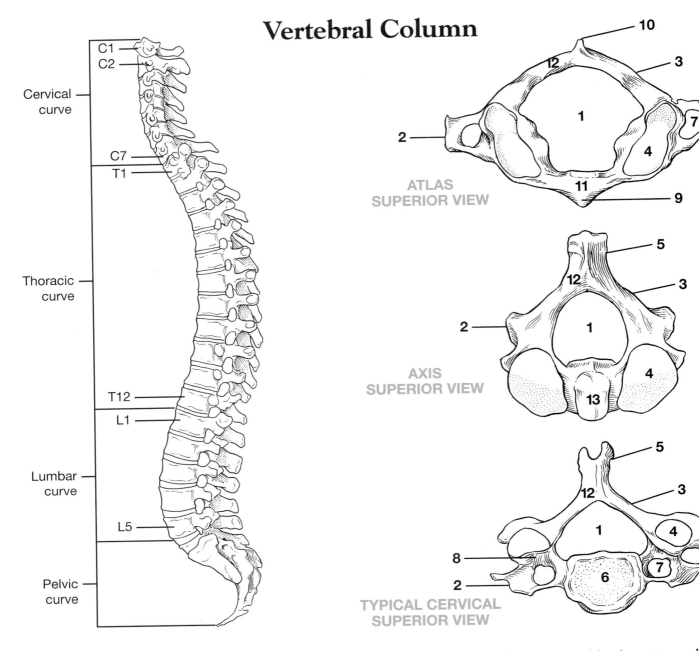

Cervical curve
C1
C2
C7
T1

Thoracic curve

T12
L1

Lumbar curve
L5

Pelvic curve

ATLAS SUPERIOR VIEW
10
12
3
2
1
7
4
11
9

AXIS SUPERIOR VIEW
5
12
3
2
1
4
13

TYPICAL CERVICAL SUPERIOR VIEW
5
12
3
4
8
2
1
7
6

1. Vertebral (spinal) foremen
2. Transverse process
3. Lamina
4. Superior articulating facet
5. Spinous process
6. Body
7. Transverse foremen
8. Pedicle
9. Anterior tubercle
10. Posterior tubercle
11. Anterior arch
12. Posterior arch
13. Odontoid process (dens)

The **odontoid process** is actually the missing "body" of the atlas. The odontoid process acts as a pivot for the rotation of the atlas allowing side-to-side rotation of the head signaling "no". The anterior longitudinal ligament is severely stretched and may be torn during severe hyperextension of the neck causing a **whiplash** injury. There may also be a hyperflexion injury as the neck snaps back to the thorax. **Facet jumping** or locking of the cervical vertebrae may occur due to the dislocation of the facets.

Vertebral Column

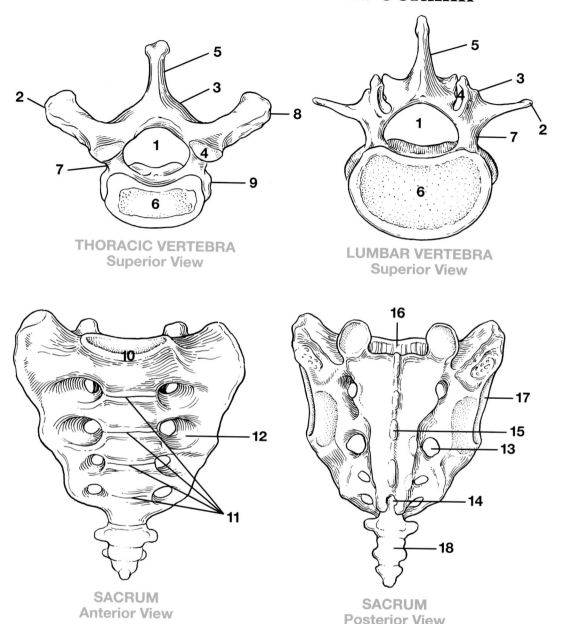

THORACIC VERTEBRA
Superior View

LUMBAR VERTEBRA
Superior View

SACRUM
Anterior View

SACRUM
Posterior View

1. Vertebral foramen
2. Transverse process
3. Lamina
4. Superior articulating facet
5. Spinous process
6. Body
7. Pedicle
8. Facet for tubercle of rib
9. Demifacet for head of rib
10. Sacral promontory
11. Transverse lines
12. Anterior sacral (pelvic) foramen
13. Posterior sacral foramen
14. Sacral hiatus
15. Medial sacral crest
16. Sacral canal
17. Sacroiliac articulating surface
18. Coccyx

A herniated or protruding intervertebral disk is often called a **slipped disk.** Protrusions of the disk usually occur posterolaterally where the supporting tissue is weak and poorly supported by the posterior ligaments. The protruding part may compress an adjacent spinal nerve root causing severe low back and leg pain. **Sciatica** is acute low back pain that radiates down the posterolateral aspect of the thigh and leg often caused by a protrusion of a lumbar intervertebral disk at the L5, S1 level. The **sacral hiatus** is a common site for an epidural injection. Lower back pain can also be caused by **facet syndrome** which involves the superior articulating facet of one vertebra and the inferior articulating facet of the adjacent vertebra. Repetitive stress and osteoarthritic changes can lead to facet hypertrophy causing pain and motion restriction and can cause narrowing of the neural foramen.

Vertebral Column Disorders and Injuries

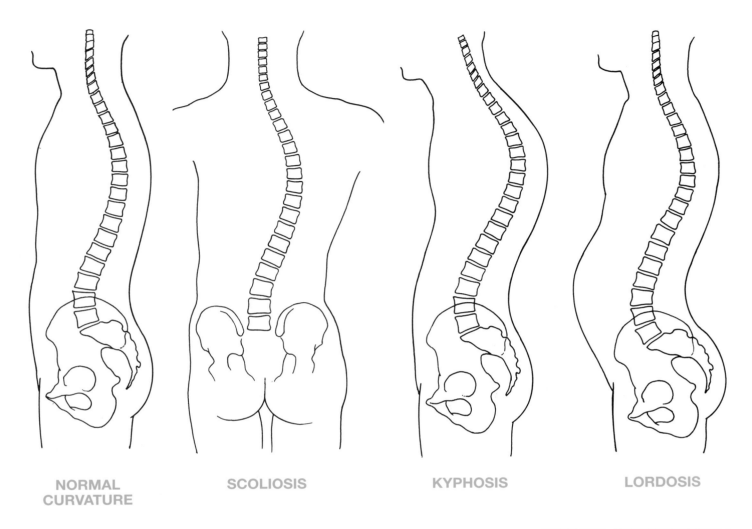

NORMAL CURVATURE   **SCOLIOSIS**   **KYPHOSIS**   **LORDOSIS**

There are four normal curvatures in the vertebral column. The thoracic and sacral are concave anteriorly, whereas the cervical and lumbar curvatures are convex anteriorly. In some people the vertebral column may show **abnormal vertebral curvatures** caused by developmental or pathological processes. **Scoliosis** is a lateral curvature of one or more vertebral segments. More common in females, it becomes apparent in the teen years due to unequal growth of the two sides of the vertebral column caused by unequal development of vertebral muscles or of both sides of the vertebrae. **Kyphosis** is an exaggeration of the normal convex curvature of the thoracic vertebrae, producing a "humpback" or "Dowager's Hump" frequently related to osteoporosis. **Lordosis** is an exaggeration of the normal convex curvature of the lumbar region, frequently referred to as "swayback". Pregnancy often develops a temporary "lumbar lordosis" and lower back pain that is usually corrected at childbirth.

Clavicle

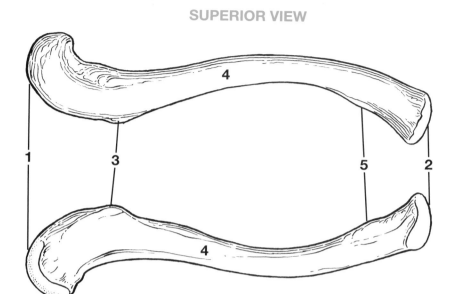

1. Acromial extremity
2. Sternal extremity

3. Conoid tubercle
4. Body of clavicle

5. Costal tuberosity

Due to its S-shaped configuration, the **clavicle** is susceptible to compressive forces caused by a blow or fall on the point of the shoulder, by a direct blow to the clavicle or by falling on an outstretched arm. Of concern in a fracture of the clavicle is the potential rupture of the subclavian vessels that lie just behind this bone. Fortunately, the natural anterior curvature of the clavicle in front of these vessels usually results in the clavicle fracturing forward than in a rearward direction. Sometimes the same type of fall instead of causing a fracture will cause an **AC joint sprain.** The **acromion process** is driven away from the **clavicle** resulting in a dislocation sometimes referred to as a separated shoulder.

Right Scapula

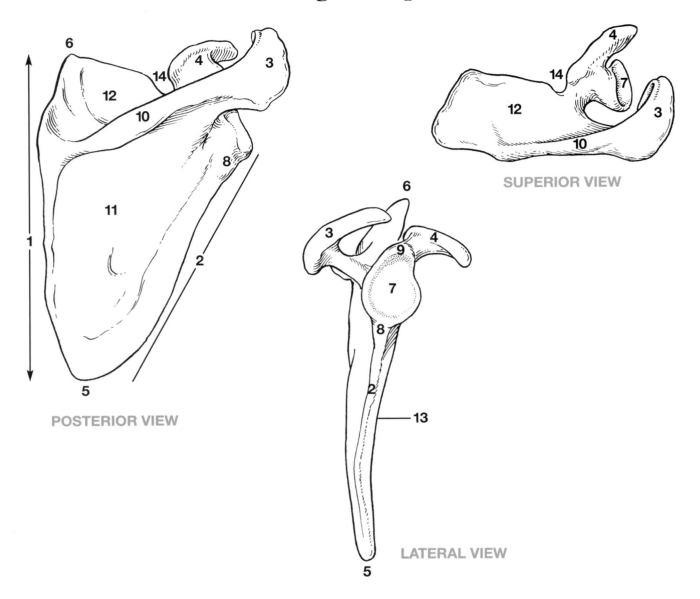

POSTERIOR VIEW

SUPERIOR VIEW

LATERAL VIEW

1. Vertebral border
2. Axillary border
3. Acromion process
4. Coracoid process
5. Inferior angle

6. Superior angle
7. Glenoid fossa
8. Infraglenoid tubercle
9. Supraglenoid tubercle

10. Spine
11. Infraspinous fossa
12. Supraspinous fossa
13. Subscapular fossa
14. Suprascapular notch

Compare the difference in depth between the **glenoid fossa** of the scapula and the **acetabulum** of the os coxa. The shallower glenoid fossa permits a greater range of motion of the arm at the shoulder, but leaves the shoulder joint more likely to dislocate. This explains why the shoulder is more easily dislocated than the hip.

Right Humerus

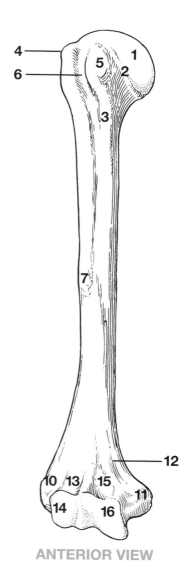

ANTERIOR VIEW

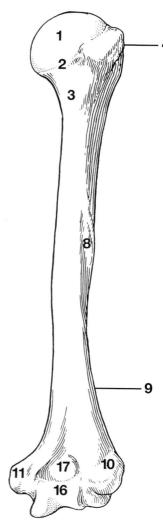

POSTERIOR VIEW

1. Head
2. Anatomical neck
3. Surgical neck
4. Greater tubercle
5. Lesser tubercle
6. Bicipital groove (intertubercular groove)
7. Deltoid tuberosity
8. Radial groove
9. Lateral supracondylar ridge
10. Lateral epicondyle
11. Medial epicondyle
12. Medial supracondylar ridge
13. Radial fossa
14. Capitulum
15. Coronoid fossa
16. Trochlea
17. Olecranon fossa

In a **greenstick fracture** the bone breaks incompletely with only one side of the bone breaking and the other bending. Greenstick fractures are common in children because their bones have relatively more organic than mineral matrix and are therefore more flexible than adult bones. This type of fracture can occur in any long bone and multiple greenstick fractures are often a sign of child abuse.

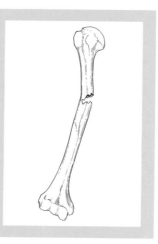

Right Radius and Ulna

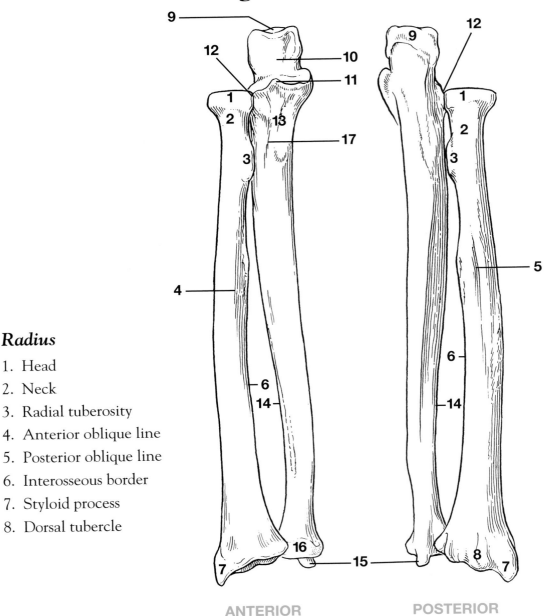

ANTERIOR

POSTERIOR

Radius

1. Head
2. Neck
3. Radial tuberosity
4. Anterior oblique line
5. Posterior oblique line
6. Interosseous border
7. Styloid process
8. Dorsal tubercle

Ulna

9. Olecranon process
10. Trochlear (semilunar) notch
11. Coronoid process
12. Radial notch
13. Ulnar tuberosity
14. Interosseous border
15. Styloid process
16. Head
17. Supinator crest

In a fall on an outstretched hand, a transverse fracture of the styloid process at the distal end of the radius may occur. This is called **Colles' Fracture**. The fractured process may be displaced posteriorly and is often forced upward or impacted into the shaft of the radius.

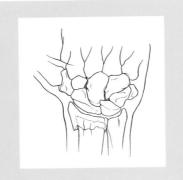

Right Hand

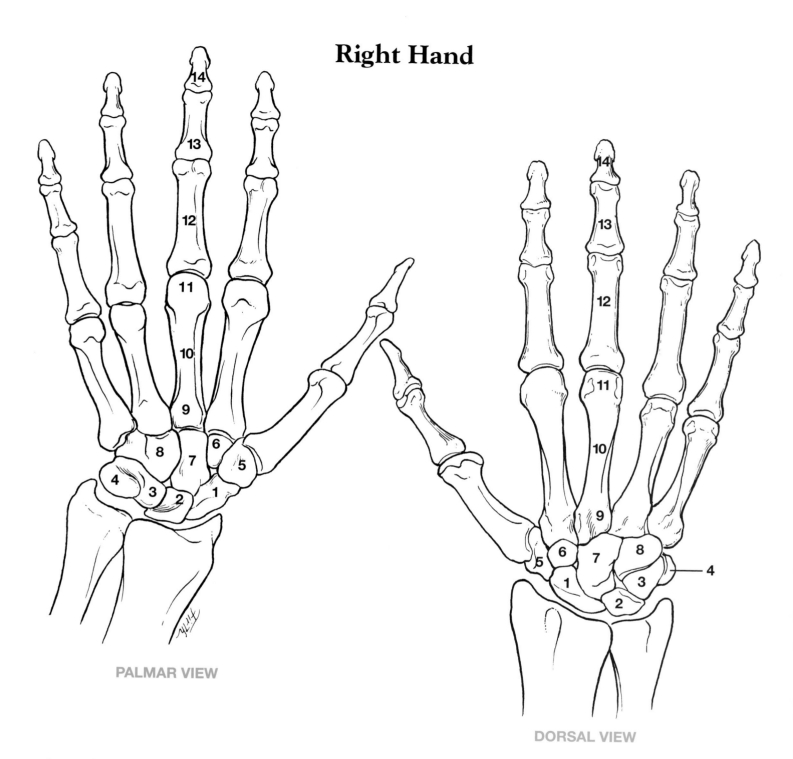

PALMAR VIEW

DORSAL VIEW

Carpals

Proximal Row

1 Scaphoid
2. Lunate
3. Triquetrum
4. Pisiform

Distal Row

5. Trapezium
6. Trapezoid
7. Capitate
8. Hamate

Metacarpals I through V

9. Base
10. Shaft
11. Head

Phalanges

12. Proximal
13. Middle
14. Distal

Os Coxae

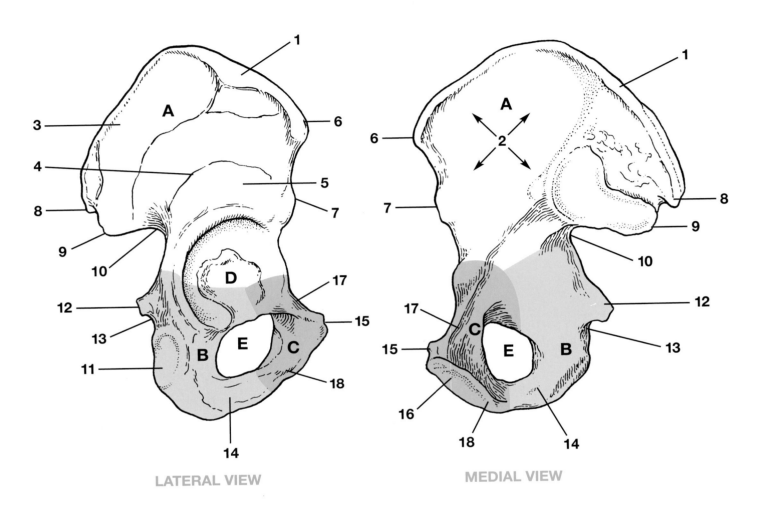

LATERAL VIEW   **MEDIAL VIEW**

A. Ilium

1. Iliac crest
2. Iliac fossa
3. Posterior gluteal line
4. Anterior gluteal line
5. Inferior gluteal line
6. Anterior superior iliac spine
7. Anterior inferior iliac spine
8. Posterior superior iliac spine
9. Posterior inferior iliac spine
10. Greater sciatic notch

B. Ischium

11. Ischial tuberosity
12. Ischial spine
13. Lesser sciatic notch
14. Ramus of ischium

C. Pubis

15. Pubic tubercle
16. Pubic symphysis
17. Superior ramus of pubis
18. Inferior ramus of pubis

D. Acetabulum

E. Obturator Foramen

Comparison of the Male and Female Pelvis

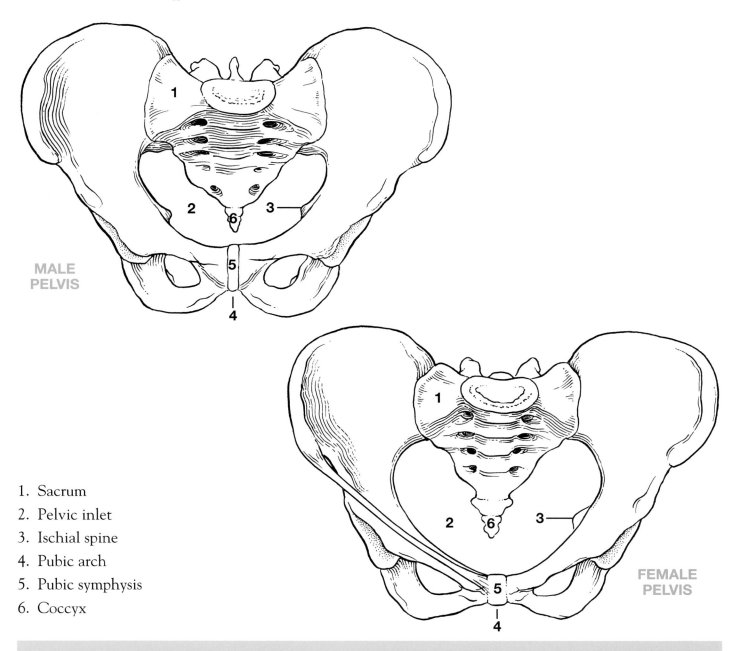

MALE
PELVIS

FEMALE
PELVIS

1. Sacrum
2. Pelvic inlet
3. Ischial spine
4. Pubic arch
5. Pubic symphysis
6. Coccyx

There are several differences in the relative size and shape of parts of the female pelvis providing a wider passageway for childbirth. These consistent differences also enable one to identify the sex of a skeleton. The female pelvis is wider, shallower and lighter. The hips (iliac crests) "flair" more in the female and the pelvic inlet is circular and proportionally larger than the heart-shaped inlet in the male. The pelvic outlet is increased in the female by a wider pubic arch (greater than 90 degrees) than in a male (less than 90 degrees). Also, in the female the **ischial tuberosities** are shorter and turned outward so the ischial spines point posteriorly. In the male the ischial tuberosities are longer and point more medially as do the ischial spines. The wider and shorter female sacrum and coccyx curve in less sharply than do the longer and more curved male sacrum and coccyx.

Right Femur

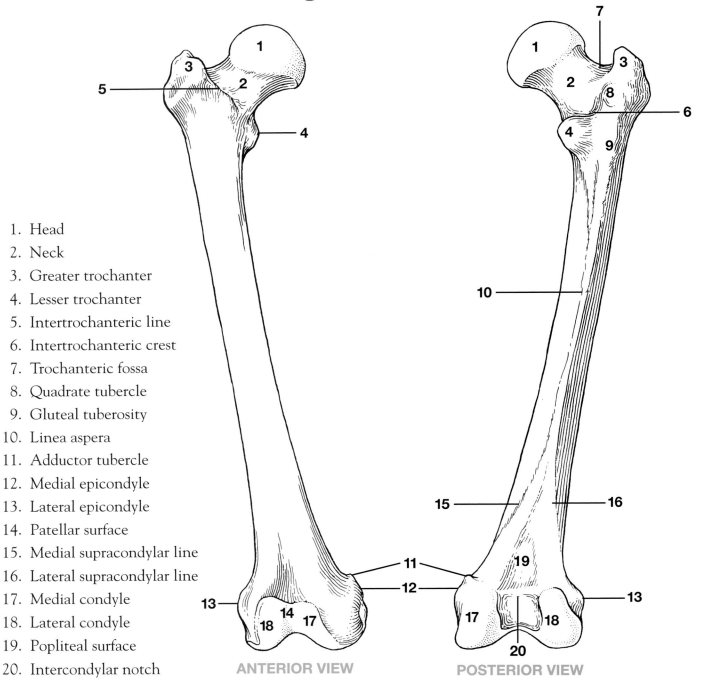

1. Head
2. Neck
3. Greater trochanter
4. Lesser trochanter
5. Intertrochanteric line
6. Intertrochanteric crest
7. Trochanteric fossa
8. Quadrate tubercle
9. Gluteal tuberosity
10. Linea aspera
11. Adductor tubercle
12. Medial epicondyle
13. Lateral epicondyle
14. Patellar surface
15. Medial supracondylar line
16. Lateral supracondylar line
17. Medial condyle
18. Lateral condyle
19. Popliteal surface
20. Intercondylar notch

ANTERIOR VIEW

POSTERIOR VIEW

Impacted fractures of the neck of the femur are common in the elderly. A severe impacted fracture may require hip replacement surgery.

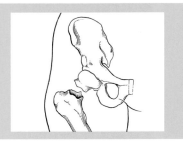

Right Tibia and Fibula

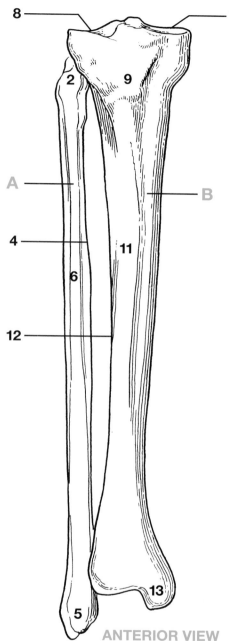

ANTERIOR VIEW

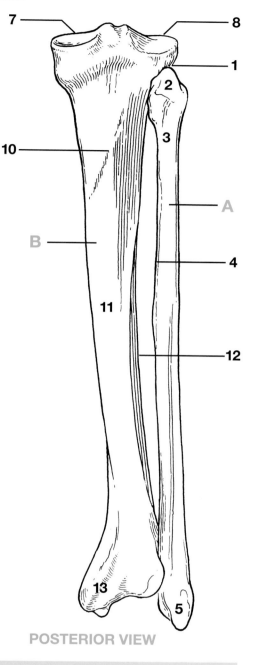

POSTERIOR VIEW

A. Fibula
1. Apex of head
2. Head
3. Neck
4. Interosseous border
5. Lateral malleolus
6. Shaft

B. Tibia
7. Medial condyle
8. Lateral condyle
9. Tibial tuberosity
10. Soleal line
11. Shaft
12. Interosseous border
13. Medial malleolus

The **fibula** is commonly fractured in skiing due to impact. It is also commonly fractured in soccer and basketball due to excessive inversion. A fracture 2 to 6 cm above the distal end of the lateral malleolus is called **Pott's fracture**. (See Chapter 3, page 49, for an illustration of Pott's fracture.) When the foot is extremely inverted, the ankle ligaments tear and the talus is forcibly tilted against the lateral malleolus, shearing it off. This fracture is relatively common in skiers and in athletes who play soccer or basketball. Because the fibula is not crucial for walking, running and jumping, it is a common source of a bone graft for reestablishing the blood supply and bone regeneration in a bone damaged due to trauma or removed because of malignancy.

Right Foot and Ankle

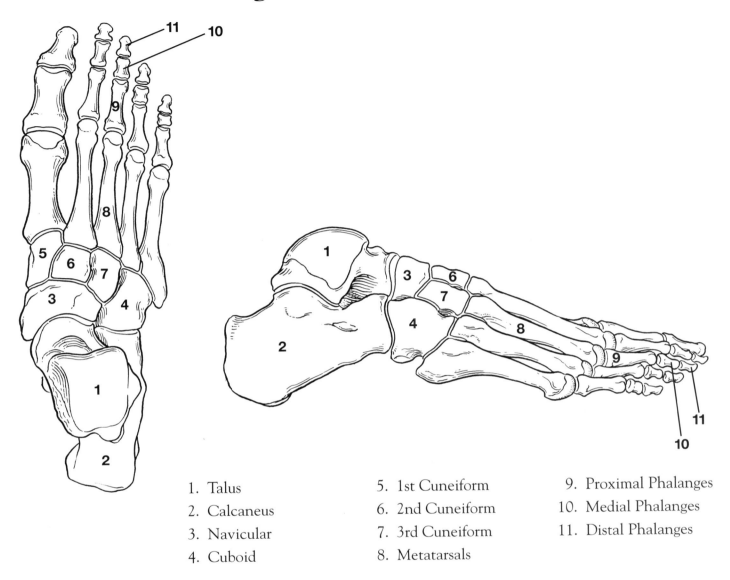

1. Talus
2. Calcaneus
3. Navicular
4. Cuboid

5. 1st Cuneiform
6. 2nd Cuneiform
7. 3rd Cuneiform
8. Metatarsals

9. Proximal Phalanges
10. Medial Phalanges
11. Distal Phalanges

Stress fractures occur in the metatarsal and navicular bones due to repetitive trauma. Runners, skaters, ballerinas are all at risk. **Bone spurs,** a build up of bone at the site of stress, frequently occur because of repetitive microtrauma. **Hammer toe** is a common deformity in which the proximal phalanx is permanently dorsiflexed at the metatarsal phalanx joint. The distal phalanx is plantar flexed giving a hammerlike appearance to the toe. **Flat feet and fallen arches** occur because during standing the plantar ligaments and fascia may be stretched beyond normal limits allowing the arch to fall. **Clubfeet** is a relatively common congenital defect in which the soles of the feet face medially and the toes point inferiorly. It can be corrected surgically. **Plantar fascitis** is the most common hind foot problem in runners. Excessive tightness of the Achilles tendon or obesity can overload the **plantar fascia's** origin on the anteromedial aspect of the calcaneus. In a chronic condition, entrapment of the first branch of the lateral planter nerve contributes to the pain which is acute with the first steps in the morning. Pain is relieved with activity but recurs after rest.

Bony Landmarks
Upper Body—Anterior View

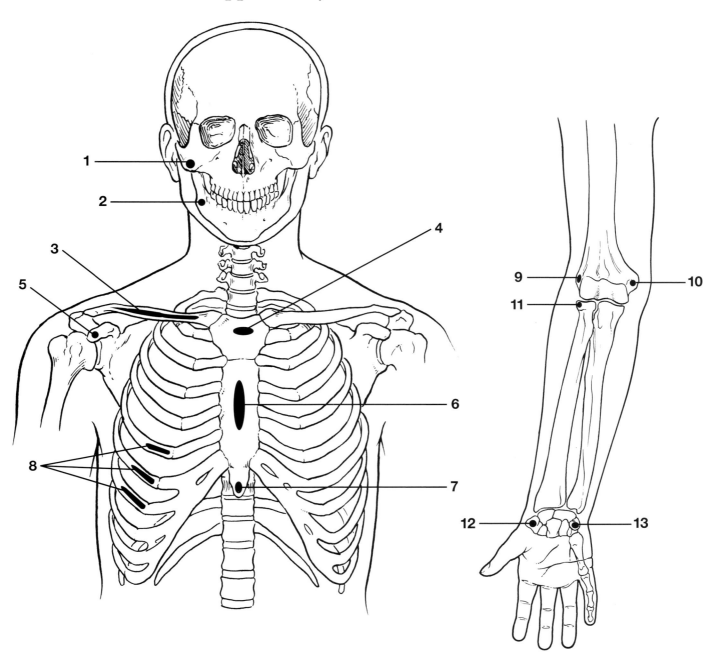

1. Zygomatic bone
2. Angle of jaw
3. Clavicle
4. Manubrium
5. Coracoid process of scapula

6. Body of sternum
7. Xiphoid process of sternum
8. Ribs
9. Lateral epicondyle of humerus
10. Medial epicondyle of humerus

11. Head of radius
12. Scaphoid bone
13. Pisiform bone

Bony Landmarks
Upper Body—Posterior View

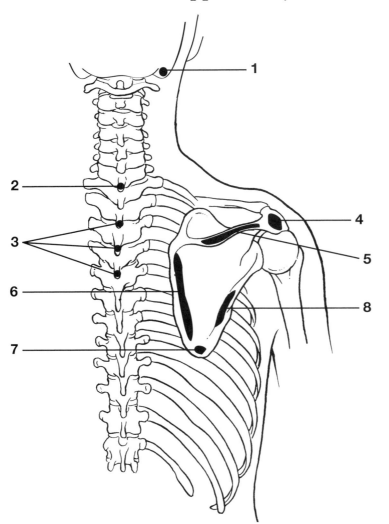
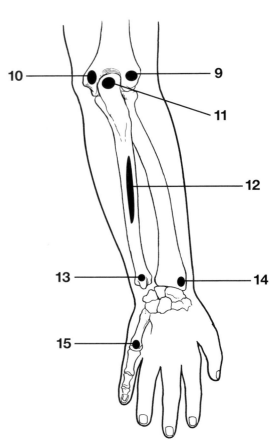

1. Mastoid process of temporal
2. Spinous process of seventh cervical vertebra
3. Spinous processes of thoracic vertebrae
4. Acromion process of scapula
5. Spine of scapula
6. Vertebral border of scapula
7. Inferior angle of scapula
8. Axillary border of scapula

9. Lateral epicondyle of humerus
10. Medial epicondyle of humerus
11. Olecranon process of ulna
12. Posterior border of ulna
13. Styloid process of ulna
14. Styloid process of radius
15. Metacarpal heads (knuckles)

Bony Landmarks
Lower Body—Anterior View

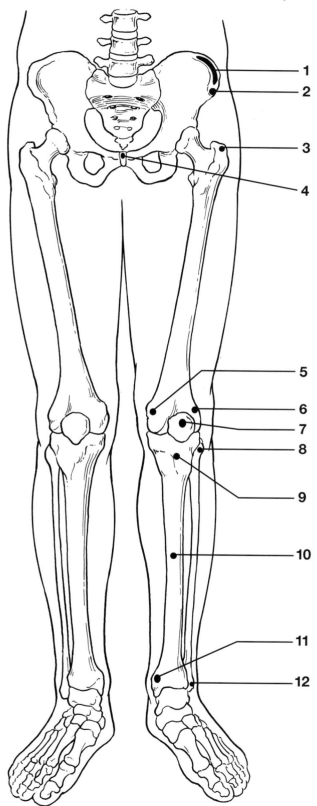

1. Iliac crest
2. Anterior superior iliac spine
3. Greater trochanter of femur
4. Pubis symphysis
5. Medial epicondyle of femur
6. Lateral epicondyle of femur
7. Patella
8. Head of fibula
9. Tuberosity of tibia
10. Anterior shaft of tibia
11. Medial malleolus of tibia
12. Lateral malleolus of fibula

Bony Landmarks
Lower Body—Posterior View

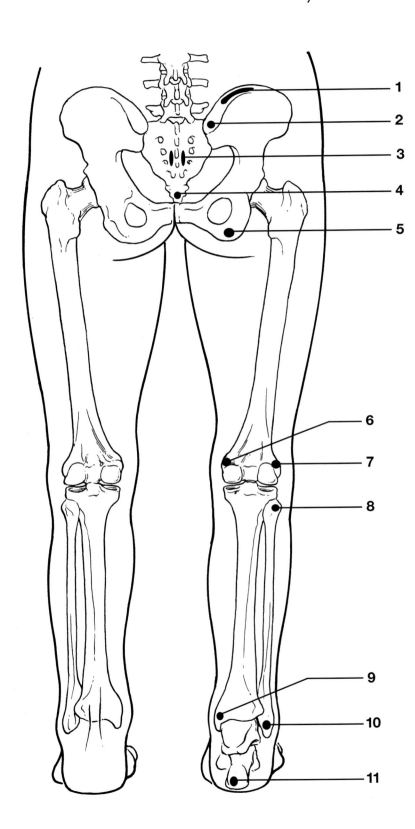

1. Iliac crest
2. Posterior superior iliac spine
3. Sacrum
4. Coccyx
5. Ischial tuberosity
6. Medial epicondyle of femur
7. Lateral epicondyle of femur
8. Head of fibula
9. Medial malleolus of tibia
10. Lateral malleolus of fibula
11. Calcaneus

THE HUMAN SYSTEMS

INDEX

THE HUMAN SYSTEMS

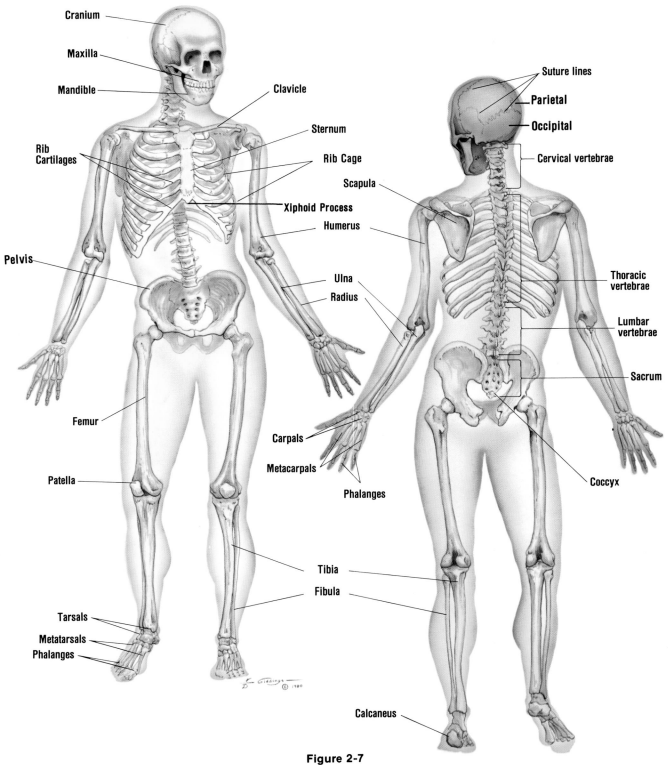

Cranium

Maxilla

Mandible

Clavicle

Sternum

Rib Cartilages

Rib Cage

Xiphoid Process

Scapula

Humerus

Pelvis

Ulna

Radius

Femur

Carpals

Metacarpals

Phalanges

Patella

Tibia

Fibula

Tarsals

Metatarsals

Phalanges

Calcaneus

Suture lines

Parietal

Occipital

Cervical vertebrae

Thoracic vertebrae

Lumbar vertebrae

Sacrum

Coccyx

D Giddings © 1980

Figure 2-7

THE SKELETAL SYSTEM

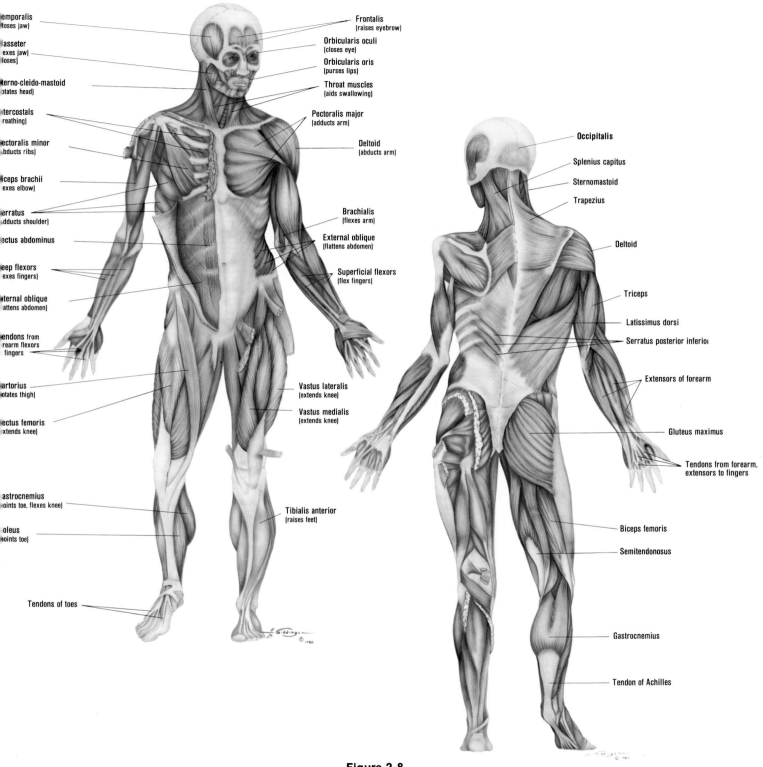

Temporalis
(closes jaw)

Masseter
(flexes jaw)
(closes)

Sterno-cleido-mastoid
(rotates head)

Intercostals
(breathing)

Pectoralis minor
(abducts ribs)

Biceps brachii
(flexes elbow)

Serratus
(adducts shoulder)

Rectus abdominus

Deep flexors
(flexes fingers)

Internal oblique
(flattens abdomen)

Tendons from
forearm flexors
to fingers

Sartorius
(rotates thigh)

Rectus femoris
(extends knee)

Gastrocnemius
(points toe, flexes knee)

Soleus
(points toe)

Tendons of toes

Frontalis
(raises eyebrow)

Orbicularis oculi
(closes eye)

Orbicularis oris
(purses lips)

Throat muscles
(aids swallowing)

Pectoralis major
(adducts arm)

Deltoid
(abducts arm)

Brachialis
(flexes arm)

External oblique
(flattens abdomen)

Superficial flexors
(flex fingers)

Vastus lateralis
(extends knee)

Vastus medialis
(extends knee)

Tibialis anterior
(raises feet)

Occipitalis

Splenius capitus

Sternomastoid

Trapezius

Deltoid

Triceps

Latissimus dorsi

Serratus posterior inferior

Extensors of forearm

Gluteus maximus

Tendons from forearm,
extensors to fingers

Biceps femoris

Semitendonosus

Gastrocnemius

Tendon of Achilles

Figure 2-8

THE MUSCULAR SYSTEM

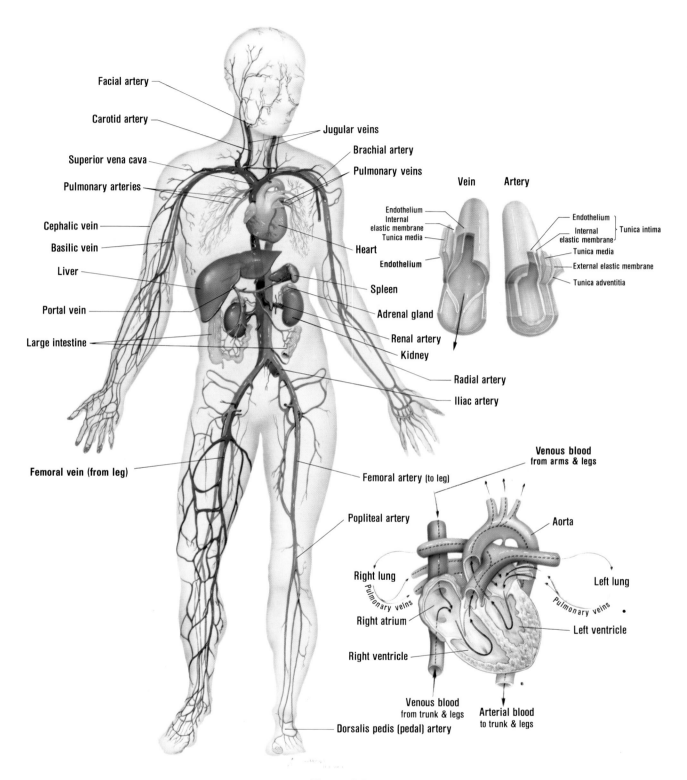

Facial artery

Carotid artery

Superior vena cava

Pulmonary arteries

Cephalic vein

Basilic vein

Liver

Portal vein

Large intestine

Femoral vein (from leg)

Jugular veins

Brachial artery

Pulmonary veins

Heart

Spleen

Adrenal gland

Renal artery

Kidney

Radial artery

Iliac artery

Femoral artery (to leg)

Popliteal artery

Right lung

Right atrium

Right ventricle

Venous blood
from trunk & legs

Dorsalis pedis (pedal) artery

Vein   Artery

Endothelium
Internal
elastic membrane
Tunica media

Endothelium

Endothelium
Internal
elastic membrane

Tunica intima

Tunica media

External elastic membrane

Tunica adventitia

Venous blood
from arms & legs

Aorta

Left lung

Pulmonary veins

Left ventricle

Arterial blood
to trunk & legs

Figure 2-9

THE CIRCULATORY SYSTEM

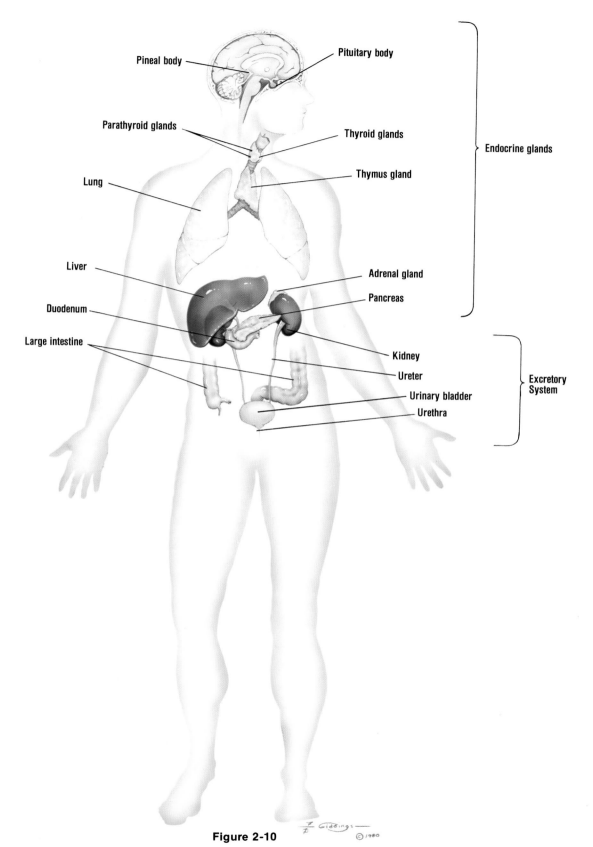

Pineal body

Pituitary body

Parathyroid glands

Thyroid glands

Thymus gland

Lung

Liver

Adrenal gland

Pancreas

Duodenum

Large intestine

Kidney

Ureter

Urinary bladder

Urethra

Endocrine glands

Excretory System

Figure 2-10

THE URINARY AND ENDOCRINE SYSTEM

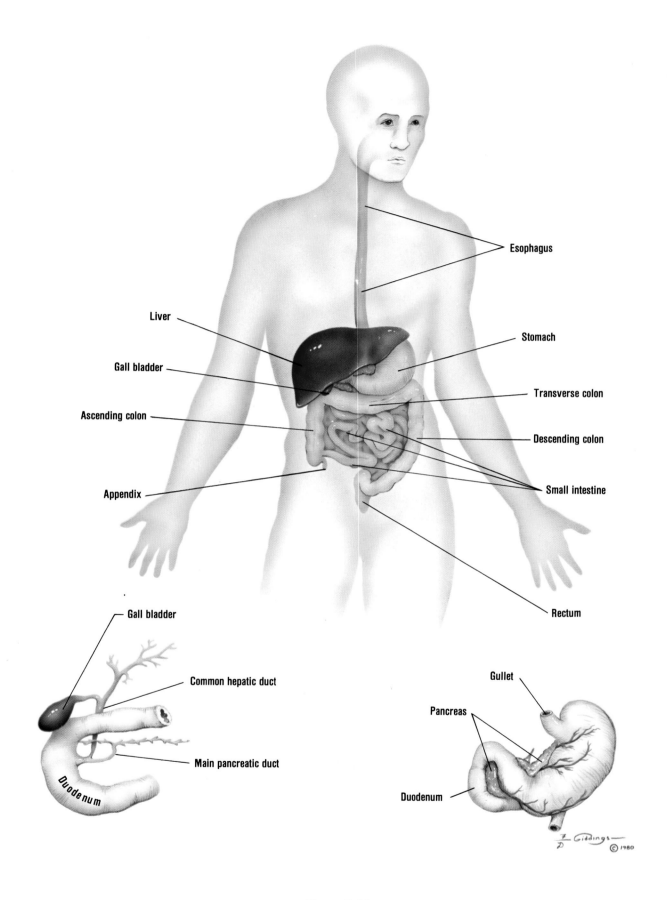

Esophagus

Liver

Gall bladder

Ascending colon

Appendix

Stomach

Transverse colon

Descending colon

Small intestine

Rectum

Gall bladder

Common hepatic duct

Main pancreatic duct

Duodenum

Gullet

Pancreas

Duodenum

Figure 2-11

THE DIGESTIVE SYSTEM

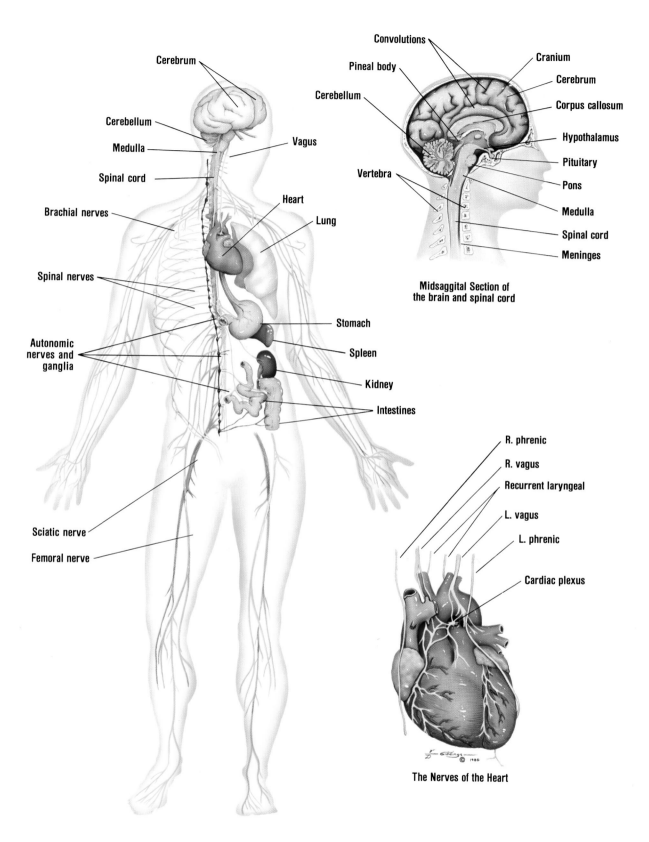

Cerebrum

Cerebellum

Medulla

Spinal cord

Vagus

Brachial nerves

Heart

Lung

Spinal nerves

Autonomic
nerves and
ganglia

Stomach

Spleen

Kidney

Intestines

Sciatic nerve

Femoral nerve

Convolutions

Pineal body

Cerebellum

Cranium

Cerebrum

Corpus callosum

Hypothalamus

Pituitary

Vertebra

Pons

Medulla

Spinal cord

Meninges

**Midsaggital Section of
the brain and spinal cord**

R. phrenic

R. vagus

Recurrent laryngeal

L. vagus

L. phrenic

Cardiac plexus

The Nerves of the Heart

Figure 2-12

THE NERVOUS SYSTEM

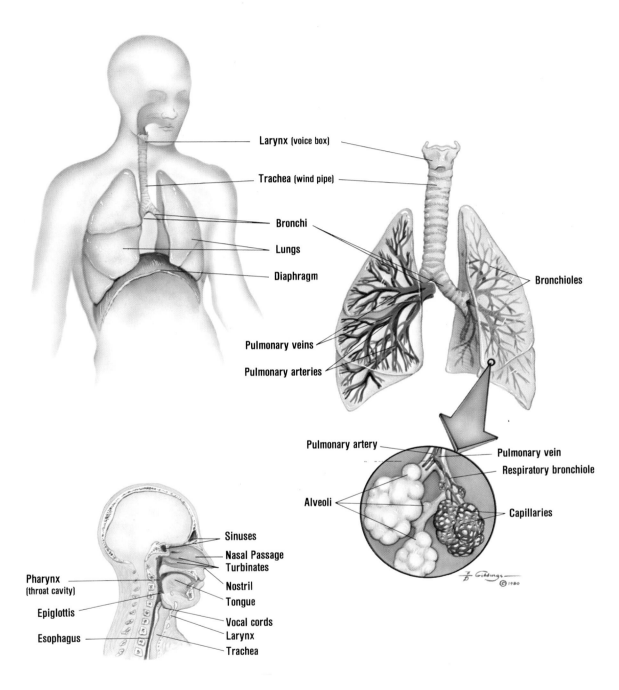

Larynx (voice box)

Trachea (wind pipe)

Bronchi

Lungs

Diaphragm

Bronchioles

Pulmonary veins

Pulmonary arteries

Pulmonary artery

Pulmonary vein

Respiratory bronchiole

Alveoli

Capillaries

Sinuses

Nasal Passage
Turbinates

Pharynx
(throat cavity)

Nostril

Tongue

Epiglottis

Vocal cords

Esophagus

Larynx

Trachea

Figure 2-13

THE RESPIRATORY SYSTEM

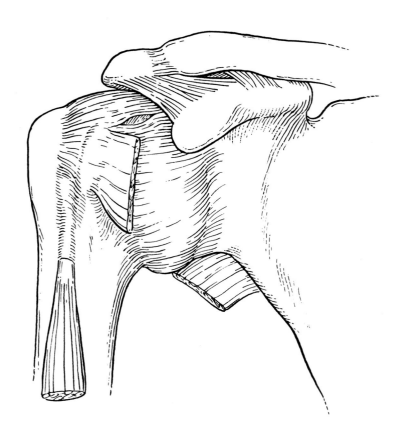

Articulation

Fibrous Articulations

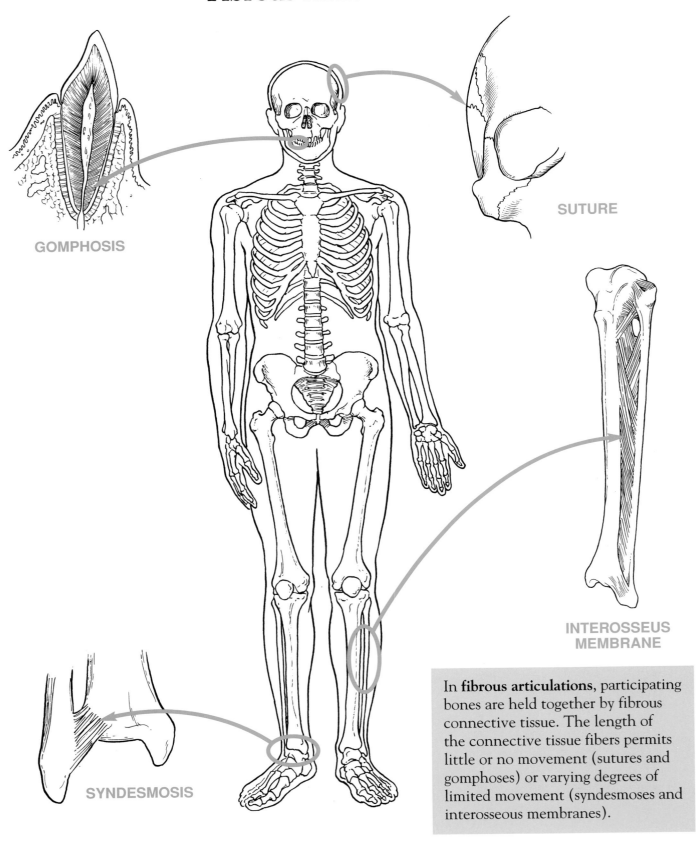

GOMPHOSIS

SUTURE

INTEROSSEUS MEMBRANE

SYNDESMOSIS

In **fibrous articulations,** participating bones are held together by fibrous connective tissue. The length of the connective tissue fibers permits little or no movement (sutures and gomphoses) or varying degrees of limited movement (syndesmoses and interosseous membranes).

Cartilaginous Articulations

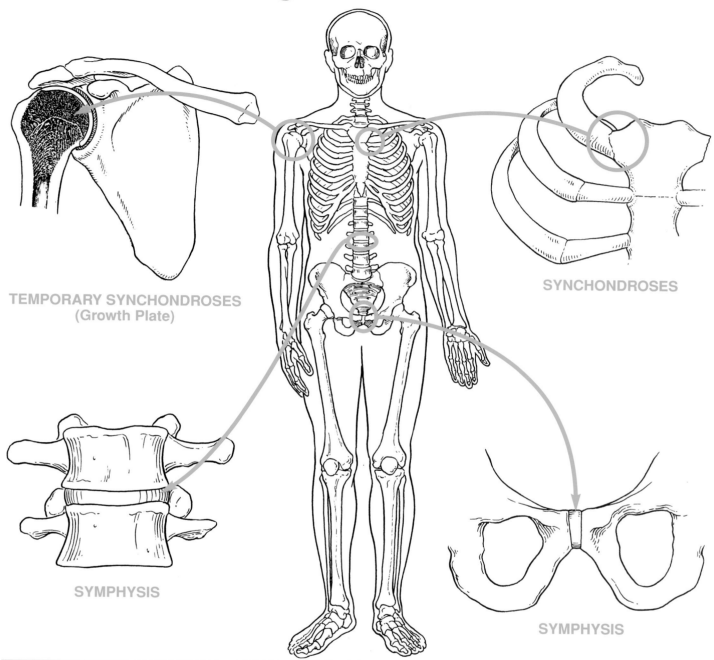

TEMPORARY SYNCHONDROSES
(Growth Plate)

SYNCHONDROSES

SYMPHYSIS

SYMPHYSIS

In these articulations bones are joined by either hyaline cartilage or fibrocartilage. In **synchondroses**, hyaline cartilage forms either temporary ("growth plates" in developing bones) or permanent (the first rib costal cartilage and the manubrium of the sternum) articulations.

Some texts still indicate that all rib/sternum joints are synchondroses, but close examination has determined that ribs 2–7 articulate with the sternum by gliding synovial (freely moveable) articulations. Fibrocartilaginous **symphyses** are slightly moveable and provide strength and flexibility between the spool-shaped bodies of vertebrae ("intervertebral discs") and between the pubic portions of the pelvis (pubic symphysis).

Synovial Articulations

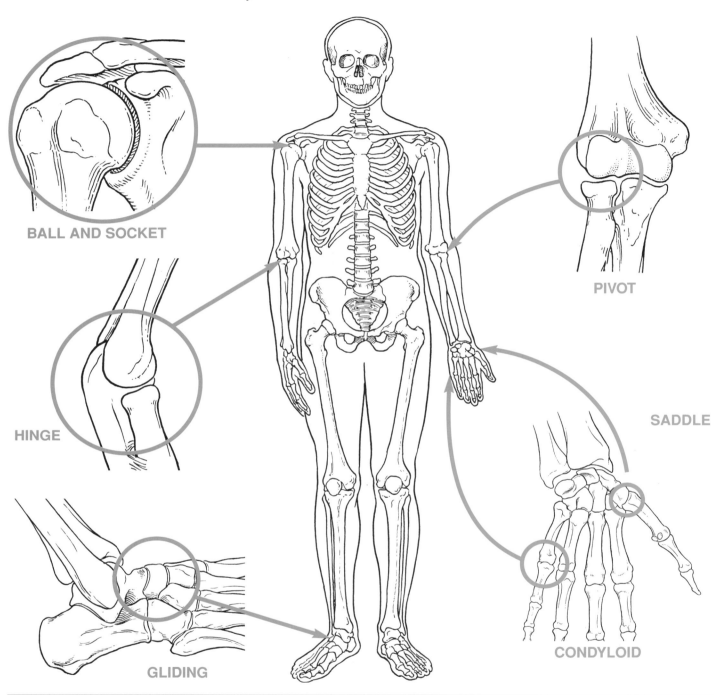

BALL AND SOCKET

PIVOT

HINGE

SADDLE

GLIDING

CONDYLOID

The majority of articulations are freely moveable **synovial joints**. The adjacent bony surfaces are capped by hyaline cartilage and joined by a fibrous articular capsule made up of an outer layer of ligaments and inner synovial membranes that produce a viscous synovial fluid that fills the synovial cavity and lubricates the joint. Additional support is provided by surrounding ligaments and tendons. The direction and magnitude of movement allowed by each of the six subkinds of synovial joints is determined by the shapes of the apposing bone surfaces.

"Typical" Synovial Joint Structure

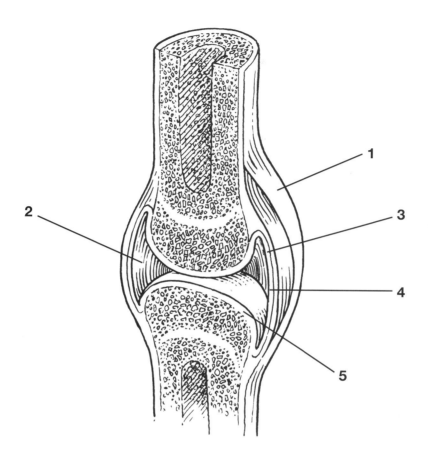

1. Ligament
2. Joint cavity
3. Fibrous capsule
4. Synovial membrane
5. Articular cartilage

The majority of articulations are freely moveable **synovial joints**. The adjacent bony surfaces are capped by hyaline cartilage and joined by a fibrous capsule made up of an outer layer of fibrous connective tissue and an inner synovial membrane that produces a viscous synovial fluid that fills the synovial cavity and lubricates the joint. Additional support is provided by surrounding ligaments and tendons. The direction(s) and amount of movement allowed by each of the six subkinds of synovial joints is determined by the shapes of the opposing bony surfaces.

Condyloid Subkind of Synovial Joints
Metacarpal-Phalangeal Joint

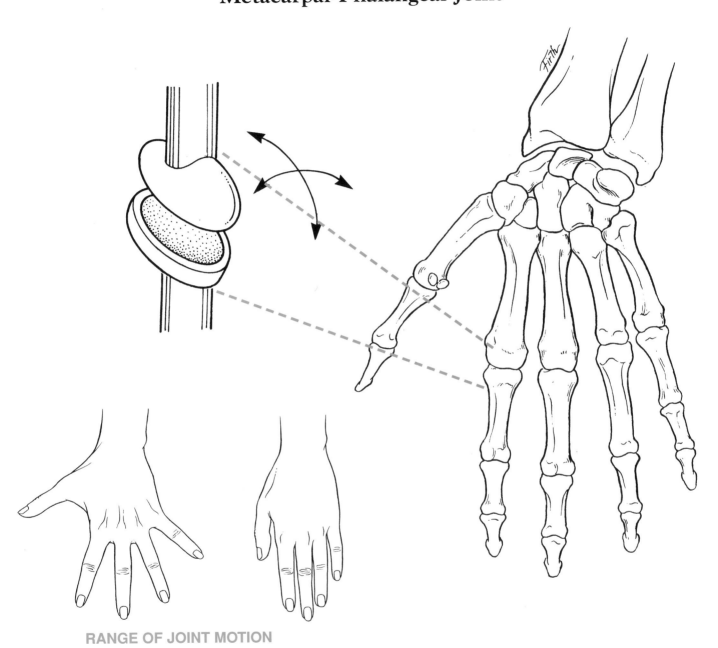

RANGE OF JOINT MOTION

Condyloid joints allow movement in two directions. Simulate the movements of the hand diagrams by flexing or bending your fingers at the knuckles; then straighten and spread your fingers sidewards. Other examples of condyloid joints are the occipital-atlas, temporal-mandibular (TMJ), radius-carpal, carpal-metacarpal, femur-tibia, tibia/fibula-talus, tarsal-metatarsal and metatarsal-phalanges. Although many of these joints are often considered hinge joints because the most pronounced movement is flexion, they do exhibit movement in another direction as well.

Rotation Subkind of Synovial Joints
Radius-Ulna Joint

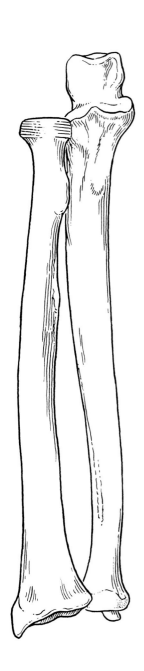

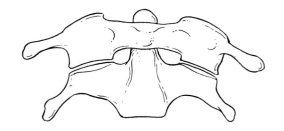

In a **pivot** or **rotational** joint, the only movement allowed is rotation around the longitudinal axis of the bone. Examples include the rotation of the **atlas** around the **odontoid process** of the **axis** and the proximal articulations between the **radius** and **ulna** as the radius rotates within the **annular ligament**.

Saddle Subkind of Synovial Joints
Thumb Carpometacarpal Joint

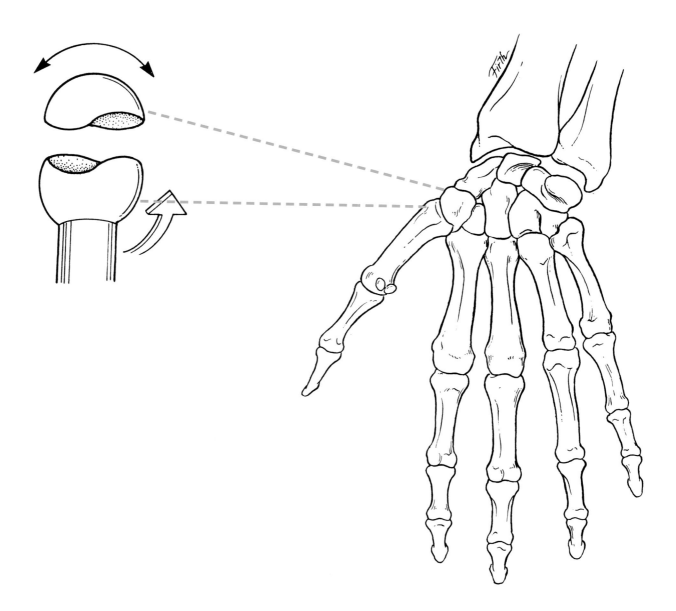

Saddle joints allow the same movements as condyloid joints—flexion, extension, abduction, adduction, and circumduction. The articular surface of each bone is concave in one direction and convex in the other. Therefore, the bones fit together as two English riding saddles would if they were rotated 90 degrees in relation to each other. The only true saddle joint is the carpometacarpal joint of the thumb.

Hinge Subkind of Synovial Joints
Humerus-Ulna Joint

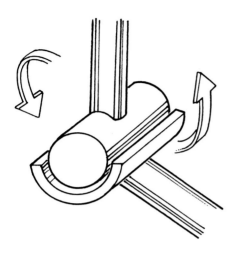

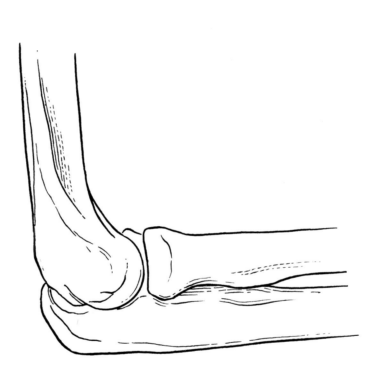

In **hinge joints**, the articular surfaces are shaped such that the only movements allowed are flexion and extension. The elbow, knee and interphalangeal joints have long been considered examples of **hinge joints**. Today, there is controversy about the placement of the knee joint in this group.

Gliding Subkind of Synovial Joints
Intercarpal and Intertarsal Joints

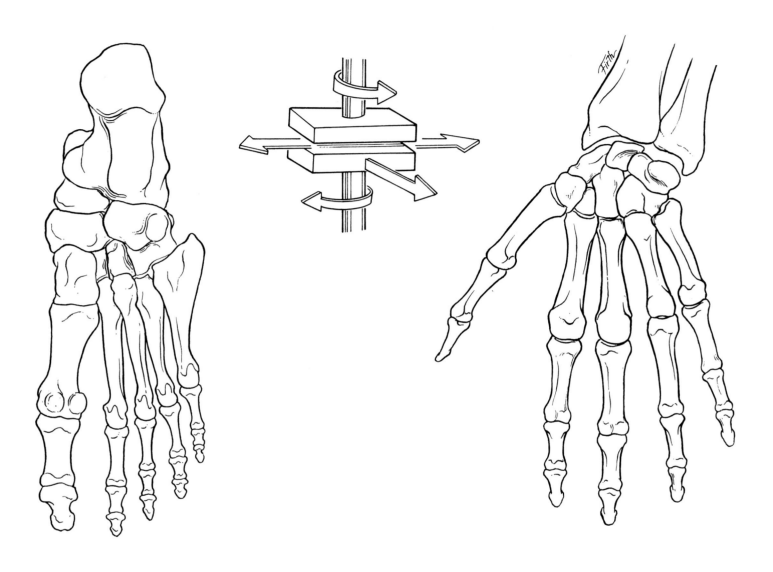

Gliding joints are also called "plane" or "planar" joints. They are one of the most numerous types of joints in the body. The articular surfaces are flat or nearly so and best illustrated by the flat superior and inferior articulating surfaces or facets on the upper thoracic vertebrae. Even though the intercarpal and intertarsal surfaces are flat, they also illustrate gliding joints. These types of freely moveable joints allow short gliding movement in many directions, not around an axis, and thus are referred to as **nonaxial** joints. Other examples include vertebrocostal, sternocostal 2–7, and sterno- and acromioclavicular joints.

Ball and Socket Subkind of Synovial Joints
Shoulder and Hip Joints

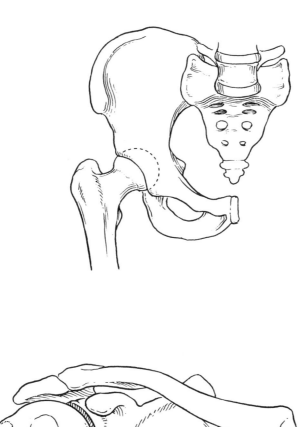

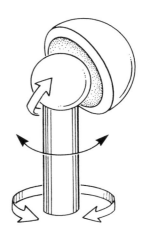

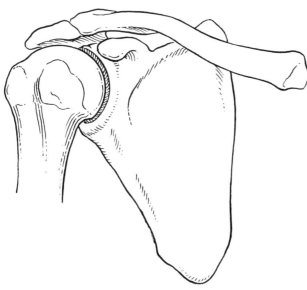

Ball and socket joints are formed by a spherical head of one bone fitting into a cup-shaped cavity on the other. Such joints allow movement around an indefinite number of axes. In addition to flexion, extension, abduction, adduction, and circumduction, ball and socket joints allow medial and lateral rotation. There are only two examples of ball and socket joints: the hip and the shoulder. The cup, acetabulum, on the hip is much deeper than that on the shoulder so that it is more difficult to dislocate. The glenoid cavity on the scapula is shallow.

Elbow Joint with Ligaments

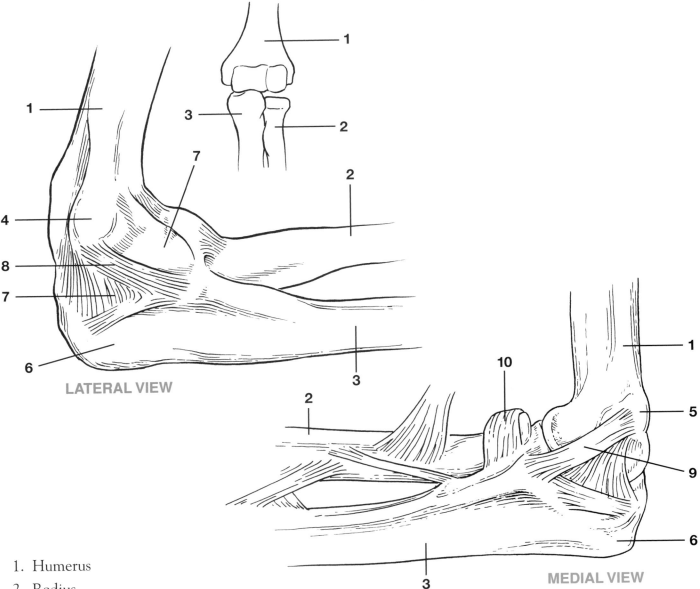

LATERAL VIEW

MEDIAL VIEW

1. Humerus
2. Radius
3. Ulna
4. Lateral condyle of the humerus
5. Medial condyle of the humerus
6. Olecranon process
7. Articular capsule
8. Radial collateral ligament
9. Ulnar collateral ligament
10. Annular ligament

The fibrous capsule completely encloses the joint. Its anterior and posterior parts are thin but its sides are strengthened by the collateral ligaments. The fibrous capsule is attached to the proximal margins of the coronoid fossa anteriorly and the olecranon fossa posteriorly. Distally the capsule is attached to the margins of the trochlear notch, the anterior border of the coronoid process and the **annular ligament**. The **radial collateral ligament** is a strong band attached proximally to the lateral epicondyle of the humerus. The **ulnar collateral ligament** is triangular. It is composed of anterior and posterior bands and is attached to the **medial epicondyle** of the humerus. The strong anterior part is attached to the tubercle on the coronoid process of the ulna. The ulnar nerve is close to the ulnar collateral ligament.

Hip Joint with Primary Ligaments

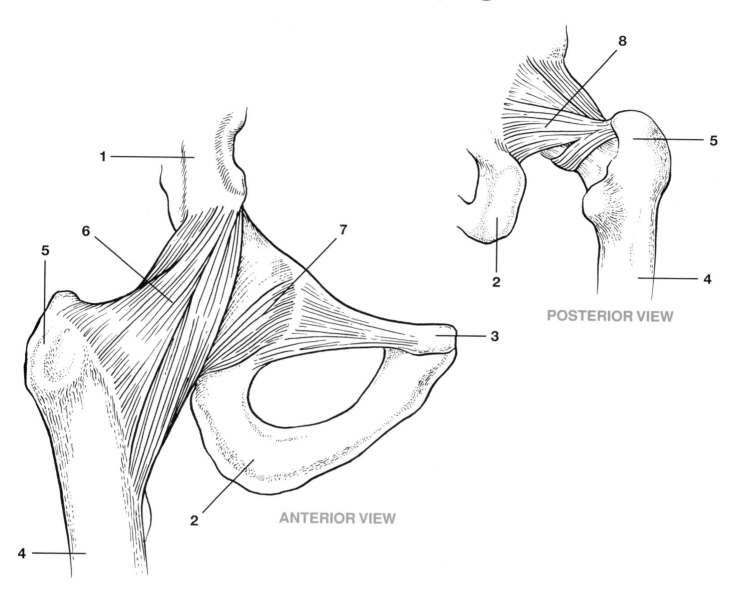

POSTERIOR VIEW

ANTERIOR VIEW

1. Ilium
2. Ischium
3. Pubis
4. Femur
5. Greater trochanter of femur
6. Iliofemoral ligament
7. Pubofemoral ligament
8. Ischiofemoral ligament

The **hip joint** is a ball and socket joint. The head of the **femur** fits into the acetabulum of the os coxa. The ligamentum teres attaches to a fovea or pit in the head of the femur. The **iliofemoral ligament** is a strong band that covers the anterior aspect of the hip joint. It attaches to the anterior inferior iliac spine and the acetabular rim. The **pubofemoral** ligament arises from the pubic part of the acetabular rim and blends with the medial part of the iliofemoral ligament. It strengthens the inferior and anterior parts of the joint. The **ischiofemoral ligament** arises from the ischial portion of the acetabular rim and spirals to the neck of the femur, medial to the base of the greater trochanter. It prevents hyperextension of the hip joint.

Knee Joint with Primary Ligaments

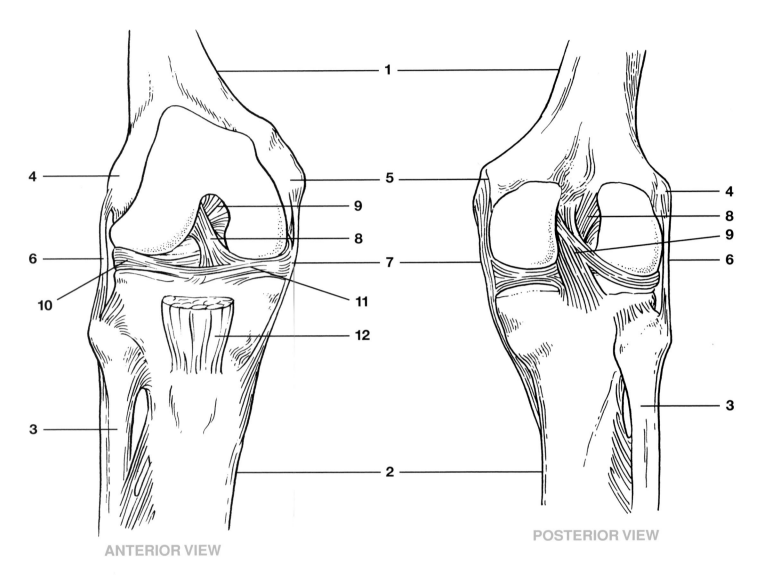

ANTERIOR VIEW

POSTERIOR VIEW

1. Femur
2. Tibia
3. Fibula
4. Lateral condyle of femur
5. Medial condyle of femur
6. Lateral (fibular) collateral ligament
7. Medial (tibial) collateral ligament
8. Anterior cruciate ligament
9. Posterior cruciate ligament
10. Lateral meniscus
11. Medial meniscus
12. Patella (tendon) ligament

The knee is the most complicated joint in the body. It is not stable as other joints, but it is one of the most often used and damaged. Standing in one position with the knees slightly flexed can cause damage to the articular cartilage. The **menisci** act as shock absorbers. If the movement against them is too abrupt, they can be crushed or torn. The two **collateral ligaments** give stability to the joint. The **anterior** and **posterior cruciate ligaments** give additional front to back stability. Tearing of the anterior cruciate ligament (ACL) is a common sports injury.

Shoulder Joint with Primary Ligaments

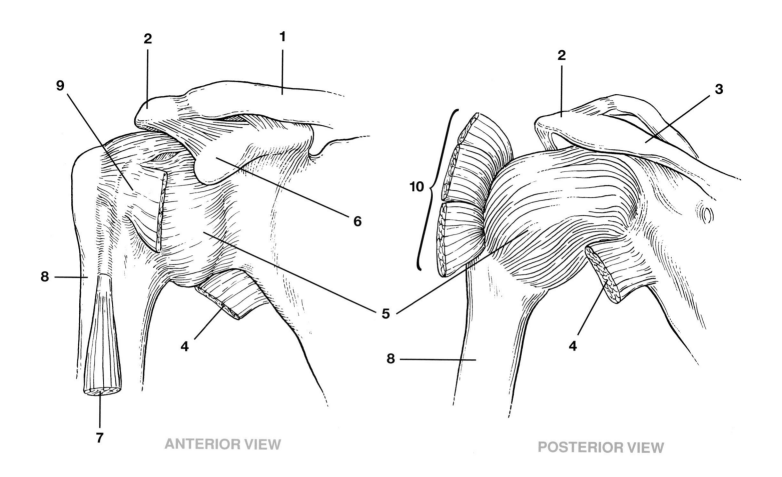

ANTERIOR VIEW

POSTERIOR VIEW

1. Clavicle
2. Acromion process
3. Spine of scapula
4. Triceps (long head)
5. Articular capsule
6. Coracoid process
7. Biceps
8. Humerus
9. Subtendinous bursa of subscapular muscle
10. Supraspinatus

The shallow glenoid fossa for the head of the humerus, together with a loose and flexible joint capsule, provide for great freedom of movement at the shoulder, but not much stability. The muscles and tendons that cross the joint provide that strength and stability, particularly the tendon of the long head of the biceps brachii and the four muscles that make up the **rotator cuff**. The rotator cuff is itself subject to stretching or tearing due to extreme movements of the arm, such as by baseball pitchers.

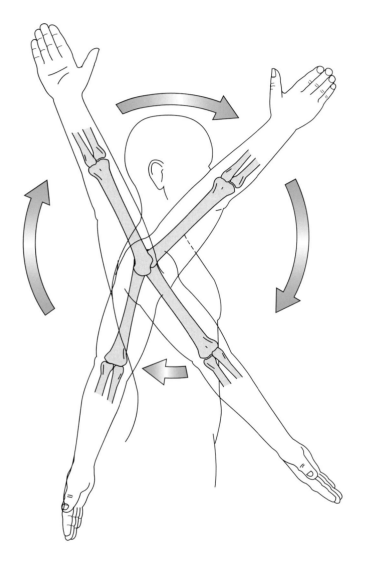

3

The Body in Motion

Flexion–Extension–Hyperextension

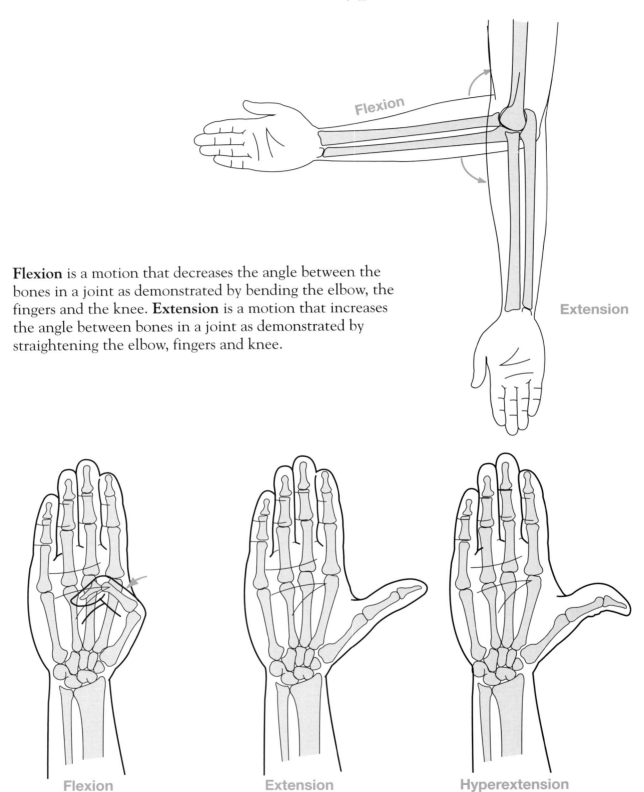

Flexion is a motion that decreases the angle between the bones in a joint as demonstrated by bending the elbow, the fingers and the knee. **Extension** is a motion that increases the angle between bones in a joint as demonstrated by straightening the elbow, fingers and knee.

Flexion

Extension

Flexion   Extension   Hyperextension

Hyperextension is a motion that goes beyond its normal limits, as seen in the "hitchhiker's thumb." Bending the head backward is an example of extension.

Abduction and Adduction

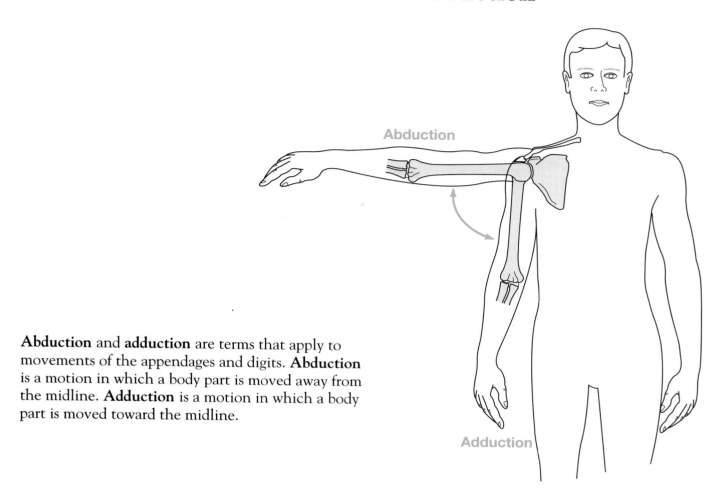

Abduction and adduction are terms that apply to movements of the appendages and digits. **Abduction** is a motion in which a body part is moved away from the midline. **Adduction** is a motion in which a body part is moved toward the midline.

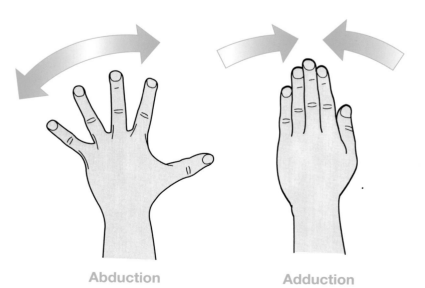

As referred to the hands and feet, abduction and adduction are defined as movement away from or toward the middle digit.

Rotation

Atlas–Axis Articulation

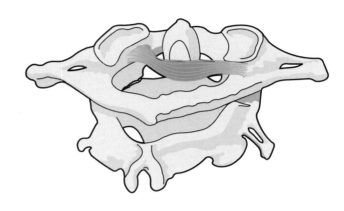

Os Coxa–Femur Lateral/Medial Rotation

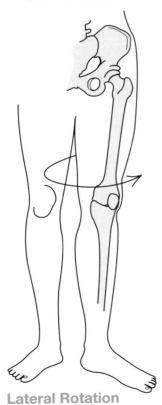
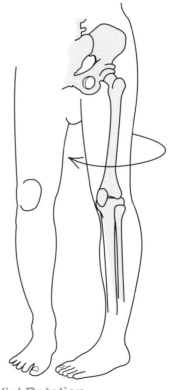

Lateral Rotation   **Medial Rotation**

Rotation is a motion in which a bone turns around its own longitudinal axis. It is the movement allowed between the first two cervical vertebrae, involving rotation of the atlas around the odontoid process (dens) of the axis, producing the characteristic side-to-side "NO" motion of the head. Medial and lateral rotation of the arm and leg occur at the hip and shoulder joints.

Supination and Pronation
Radius–Ulna Articulations

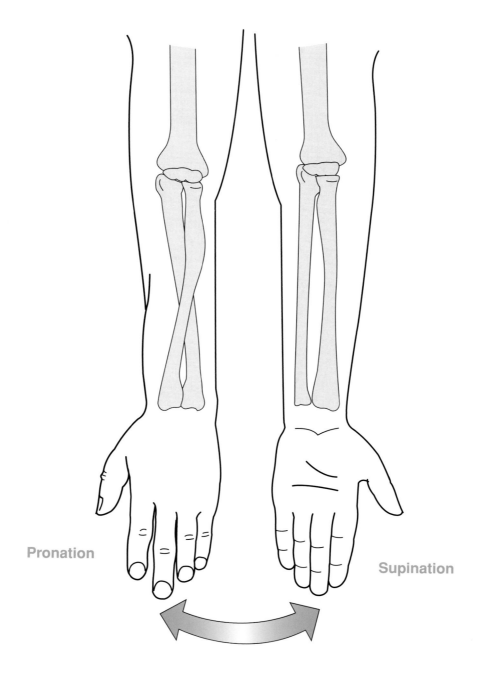

Pronation

Supination

Pronation and **Supination** are special rotational movements of the forearm that respectively cause the palm of the hand to face posterior or anterior (anatomical position). These motions are permitted by rotation between the radius and ulna at both the proximal and distal ends of the two bones. In anatomical position (supination) the two lower arm bones are parallel; in pronation the radius lies diagonally across the ulna.

Circumduction
Scapular–Humerus (shoulder) Joint

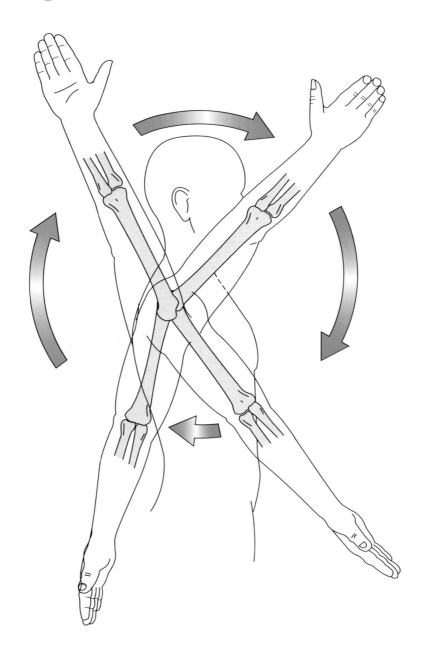

Circumduction is a motion that occurs at the "ball and socket" joints at the shoulder and the hip. In circumduction, the movement of the limbs describes a cone in space. The distal end of the limb moves in a circle while the proximal end is more or less stationary. The 360° rotation of circumduction includes flexion, abduction, extension, and rotation and is the quickest way to exercise the many muscles that move the hip and shoulder. A pitcher winding up to throw a baseball is circumducting the pitching arm.

Dorsiflexion—Plantar Flexion

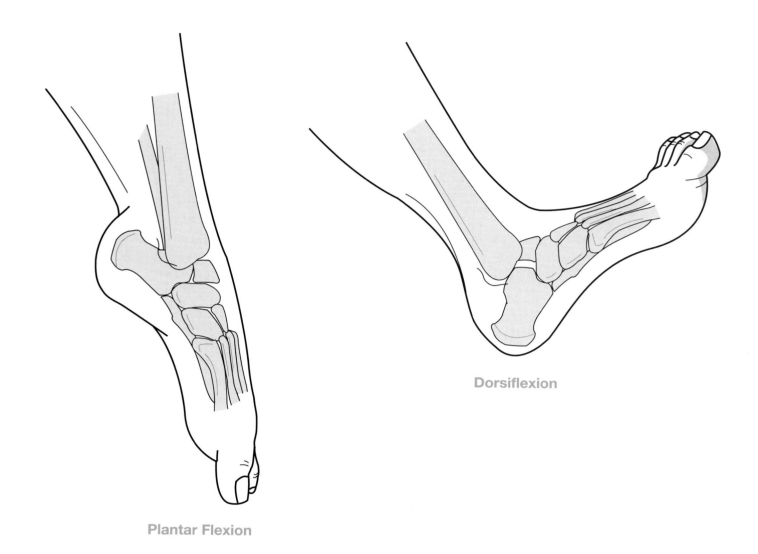

Dorsiflexion

Plantar Flexion

Plantar flexion is a downward movement of the foot and toes at the ankle resulting in the foot and toes pointing toward the floor, such as in the "toe pointed" position used by gymnasts. In older textbooks and lab manuals, plantar flexion was referred to as plantar extension. **Dorsiflexion** is an upward movement of the foot and toes at the ankle resulting in the toes and foot projecting away from the floor.

Opposition—Reposition
Thumb–Finger Touching

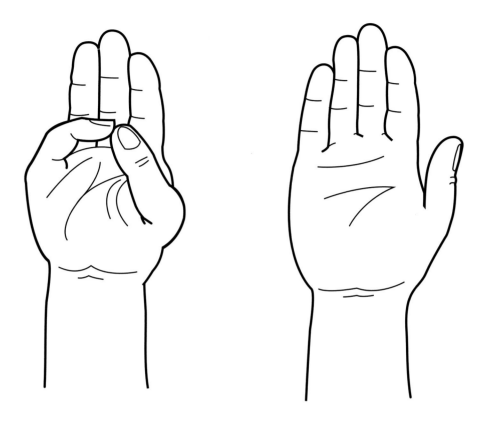

Opposition is movement of the thumb to approach or touch one or more of the fingertips; **reposition** is the reverse movement, returning the thumb to a parallel position with the fingers.

Since opposition is the movement that enables the hand to grasp objects, it is the single most important hand function. It is the motion to repair, retrain and maintain in the case of an accident.

Inversion—Eversion
Tibia/Fibula–Talus Articulation

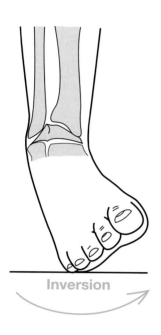
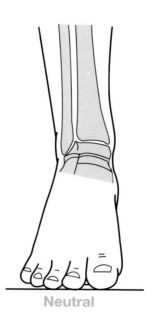
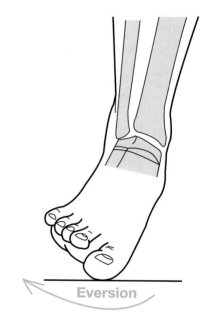

Inversion — Neutral — Eversion

Inversion and **eversion** are special movements of the foot. In inversion, the sole of the foot turns inward toward the median line; in eversion the sole of the foot turns outwards or away from the median line.

The ankle joint is the most frequently injured major articulation. **Pott's fracture** (A) is a result of forced eversion of the foot resulting in fracture of the medial malleolus with the talus shifting laterally, shearing off the lateral malleolus or the fibula superior to the inferior tibiofibular joint. Forced inversion of the foot may cause fracture of the fibular malleolus at the level of the tibial-talus joint (B).

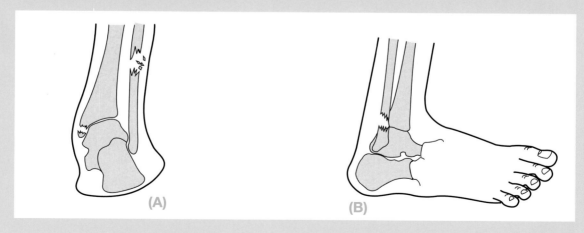

(A)   (B)

Protraction—Retraction
Temporo-Mandibular (TMJ) Articulation

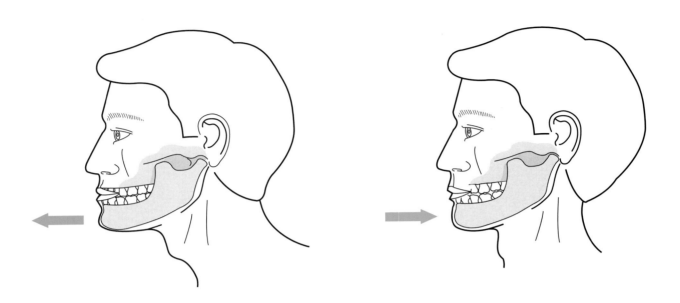

Protraction is movement of a bone anteriorly (forward) in a transverse plane; **retraction** is movement posteriorly (rearward) along a transverse plane. These are best illustrated in forward and rearward movements of the lower jaw. Similar movements of the shoulder and arms may be made through protraction and retraction of the clavicle and scapula.

Temporo-mandibular joint (TMJ) syndrome is a problem of this joint that has many symptoms, many potential causes and an equal number of possible treatments. It is characterized by one or more of the following: dull pain around the ear, tenderness of the jaw muscles, clicking or popping noise when opening or closing the mouth, limited or abnormal opening of the mouth, headache, tooth sensitivity, and abnormal wearing of the teeth. The condition may be caused by misalignment of the teeth, missing teeth, poor bite, trauma to the jaw, or arthritis, as well as anxiety, tension, clenching, bruxism, or gum chewing.

Elevation—Depression
Temporo-Mandibular (TMJ) Articulation

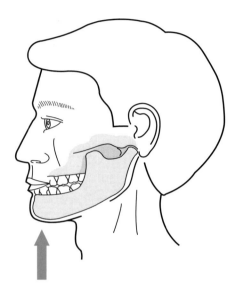 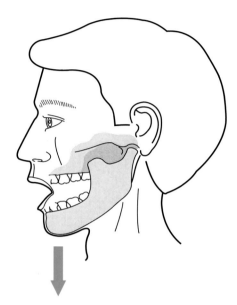

Elevation is a movement of a bone vertically such as upward movement of the clavicles when "shrugging the shoulders" when gesturing "I don't know". Elevating the lower jaw closes the mouth. **Depression** is the opposite motion, lowering the clavicles or lowering the lower jaw (opening the mouth).

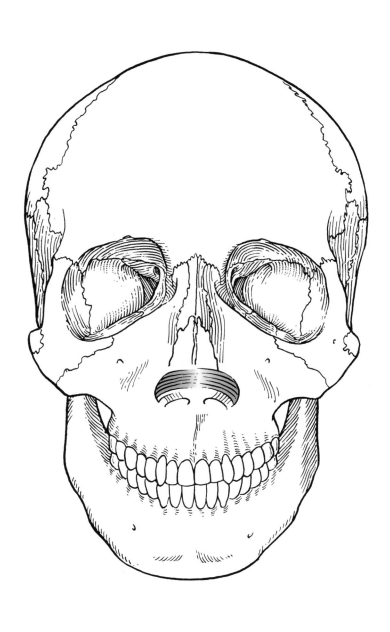

Muscles of the Face and Head

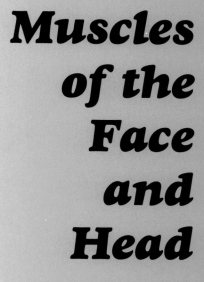

Occipital Frontalis

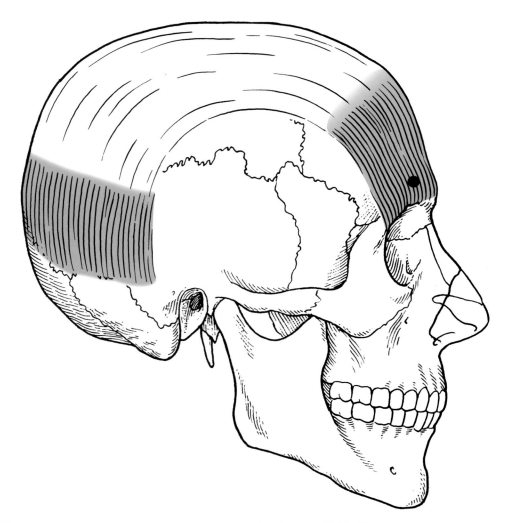

Occipital Belly

Origin:	Lateral two-thirds of superior nuchal line of occipital and mastoid process of temporal
Insertion:	Galea aponeurotica covering skull
Action:	Draws back scalp and aids in wrinkling forehead
Innervation:	Posterior auricular branch of facial nerve (VII)

Frontal Belly

Origin:	Galea aponeurotica
Insertion:	Fascia of facial muscles and skin above nose and eyes
Action:	Draws back scalp, wrinkles forehead, raises eyebrows
Innervation:	Temporal branch of facial nerve (VII)

The divisions of the **occipital frontalis** muscle are connected by a cranial aponeurosis, the galea aponeurotica. The alternate actions of these two muscles pull the scalp forward and backward and assist in wrinkling the forehead and raising the eyebrows. The **trigger point** for the frontalis portion of this muscle is the area above the eyebrow.

Temporoparietalis

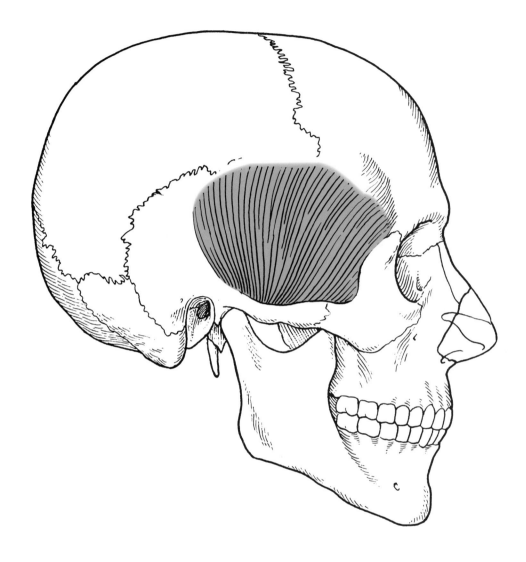

Origin: Lateral border of galea aponeurotica

Insertion: Fascia above and cartilage of the auricle

Action: Raises ears and tightens scalp

Innervation: Temporal branch of facial nerve (VII)

This part of the **epicranius** muscle is superficial to the temporalis. It is vestigial in many people but some people can use it to raise their ears.

Orbicularis Oculi

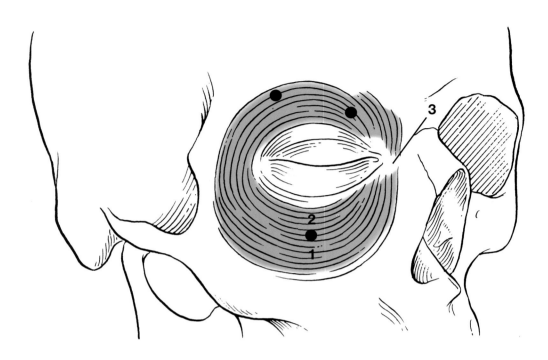

1. Orbital Part

Origin: Frontal bone and maxilla at medial margin of orbit and palpebral ligament

Insertion: Same as origin

Action: Closure of eyelids

Innervation: Temporal and zygomatic branches of facial nerve (VII)

2. Palpebral Part

Origin: Medial palpebral ligament

Insertion: Lateral palpebral ligament and zygomatic bone

Action: Closure of eyelid

Innervation: Temporal and zygomatic branches of facial nerve (VII)

3. Lacrimal Part

Origin: Lacrimal bone

Insertion: Lateral palpebral raphe

Action: Draws lacrimal canals onto surface of eye

Innervation: Temporal and zygomatic branches of facial nerve (VII)

The **orbicularis oculi** protects the eyes from intense light and injury. The various parts can be activated individually. It produces blinking, winking and squinting actions and draws the eyebrows inferiorly. Its **trigger points** are in the superior and inferior orbital areas above and below the eye. Its **referred pain pattern** is to the nose.

Depressor Septi

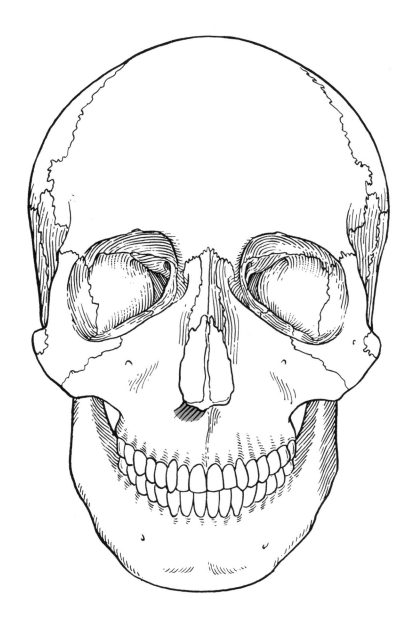

Origin: Incisive fossa of palatine process of maxilla

Insertion: Nasal septum and alar

Action: Widens nares (nostrils)

Innervation: Buccal branch of facial nerve (VII)

Orbicularis Oris

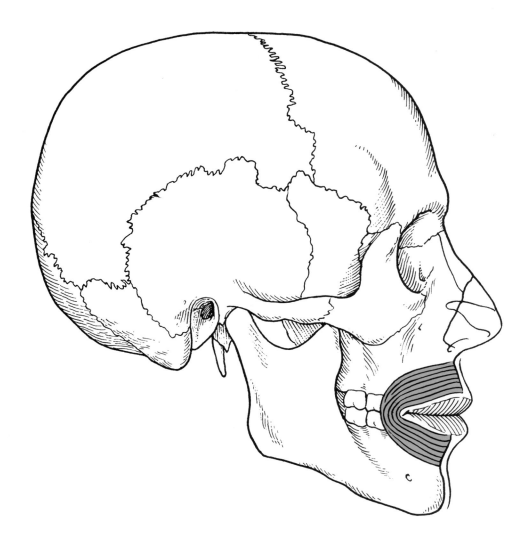

Origin: Arises indirectly from the maxilla and mandible; fibers blend with fibers of other facial muscles associated with the lips

Insertion: Encircles mouth and inserts into muscle and skin at angles of mouth

Action: Closes and protrudes lips; compresses lips against teeth

Innervation: Buccal and mandibular branches of facial nerve (VII)

This is the **kissing muscle**. It is important in whistling and in forming many letters during speech.

Levator Labii Superioris

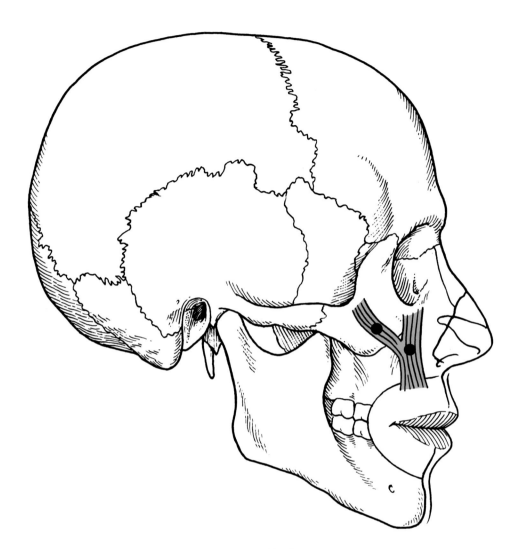

Angular head

Origin: Zygomatic bone and infraorbital margin of maxilla

Insertion: Skin and orbicularis oris muscle of upper lip

Action: Elevates upper lip and forms nasolabial furrow

Innervation: Buccal branch of facial nerve (VII)

Infraorbital head

Origin: Infraorbital margin of maxilla

Insertion: Skin and orbicularis oris muscle of upper lip

Action: Elevates upper lip

Innervation: Buccal branch of facial nerve (VII)

This thin muscle between the orbicularis oris and the eye margin raises the lip and forms the nasolabial furrow. It is used in "curling" the lip. Its **trigger points** are in the belly of the muscle. Its **referred pain pattern** is below the eye and to the bridge of the nose. It is often associated with sinus pain.

Levator Anguli Oris

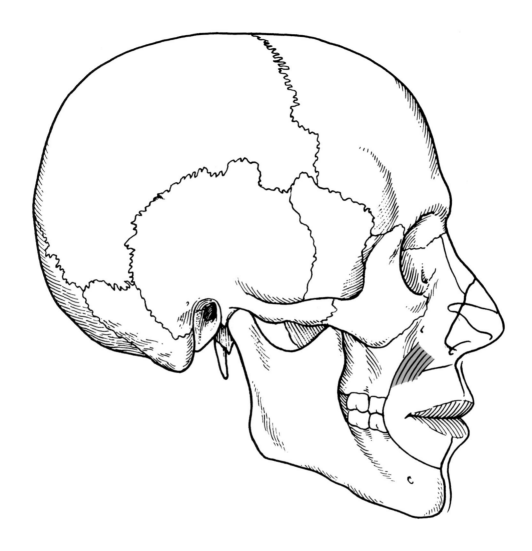

Origin: Canine fossa of maxilla

Insertion: Angle of mouth, blending with fibers of zygomaticus major, depressor anguli oris and orbicularis oris muscles

Action: Elevates corners of mouth

Innervation: Buccal branch of facial nerve (VII)

The **levator anguli oris** acts synergistically with the zygomaticus major in causing an expression of "smiling". It is also important in producing the nasolabial furrow.

Zygomaticus Major

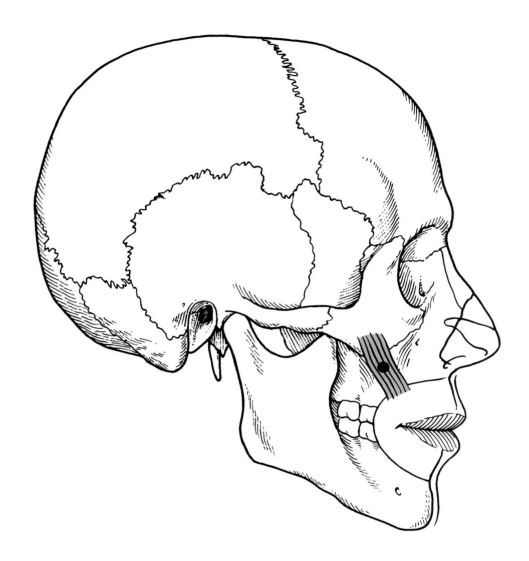

Origin: Zygomatic bone

Insertion: Angle of mouth blending with the levator anguli oris, obrbicularus oris and depressor anguli oris muscles

Action: Draws angle of mouth upward and outward

Innervation: Buccal branch of facial nerve (VII)

This is the major muscle used in smiling and laughing. The **trigger point** is in the belly of the muscle. Its **referred pain pattern** is below the eye and the side of the nose.

Zygomaticus Minor

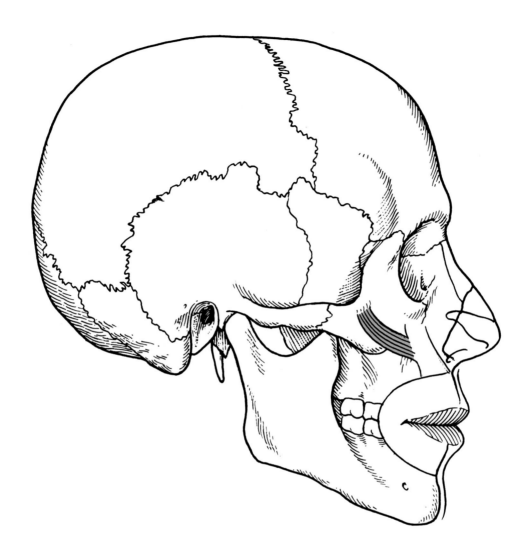

Origin: Zygomatic bone

Insertion: Angle of mouth blending with the levator anguli oris, obrbicularus oris and depressor anguli oris muscles

Action: Elevates upper lip and produces nasolabial furrow

Innervation: Buccal branch of facial nerve (VII)

Risorius

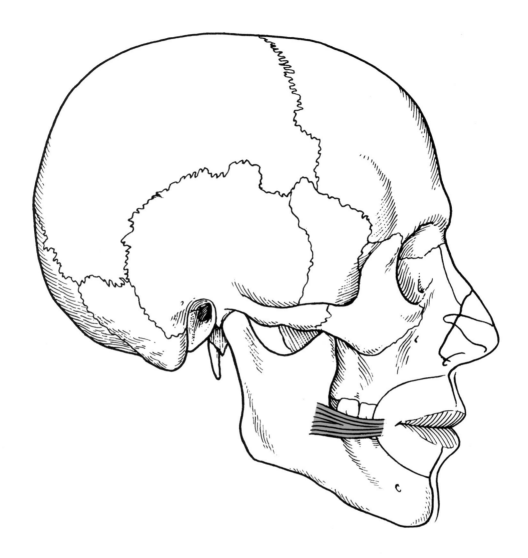

Origin: Lateral fascia over masseter muscle and parotid gland

Insertion: Skin at angle of the mouth

Action: Draws angle of the mouth laterally

Innervation: Buccal branch of facial nerve (VII)

This muscle works synergistically with the **zygomaticus major**. It tenses the mouth and draws the lips into the grinning expression.

Depressor Labii Inferioris

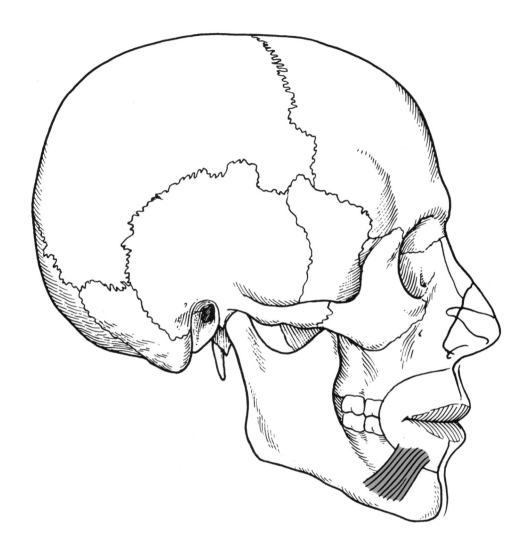

Origin: Body of the mandible lateral to the midline, between the mandibular symphysis and the mental foramen

Insertion: Skin and muscle of lower lip and blends with fibers of orbicularis oris

Action: Draws lower lip inferiorly and laterally during mastication

Innervation: Mandibular branch of facial nerve (VII)

As this muscle draws the lower lip inferiorly and laterally, it produces the typical expression of a "pout".

Depressor Anguli Oris

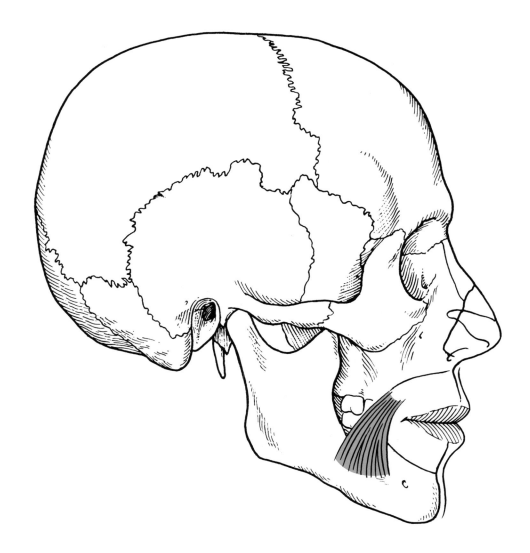

Origin: Oblique line of mandible

Insertion: Skin and muscle at angle of mouth below insertion of zygomaticus

Action: Draws angle of mouth downward and laterally

Innervation: Mandibular branch of facial nerve (VII)

This muscle draws the corners of the mouth downward and laterally in opening the mouth, as in the "Tragedy Mask" grimace. It is also used in frowning, showing disapproval and expressing the "down-in-the-mouth" grimace.

Mentalis

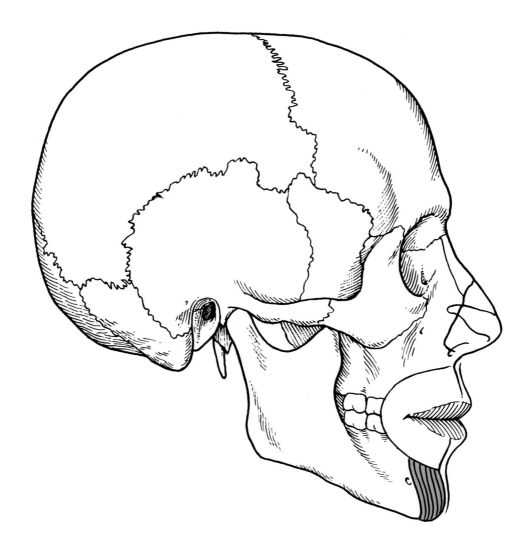

Origin: Incisive fossa of mandible

Insertion: Skin of chin

Action: Elevates and protrudes lower lip, at same time wrinkling the skin of the chin

Innervation: Mandibular branch of facial nerve (VII)

This broad muscle forms the muscle mass of the chin. It is used in "pouting".

Buccinator

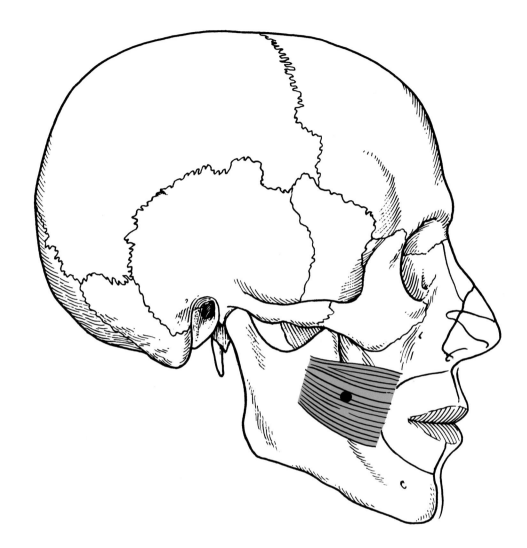

Origin: Outer surface of alveolar processes of maxilla and mandible and pterygomandibular raphe

Insertion: Angle of mouth blending with fibers of the Orbicularis oris muscle

Action: Draws corner of mouth laterally, compresses cheek

Innervation: Lower buccal branches of the facial nerve (VII)

These muscles are important in compressing the cheeks as in blowing air out of mouth. They also hold food between the teeth in chewing and cause the cheeks to cave in producing the sucking action in drinking through a straw or in a nursing infant. It is well-developed in nursing infants. When it is paralyzed, as in **Bell's palsy**, food accumulates in the oral vestibule. Bell's palsy is a unilateral paralysis of the facial muscles caused by dysfunction of cranial nerve VII. The **trigger point** is in the belly of the muscle. The **referred pain pattern** is to the upper gum.

Temporalis

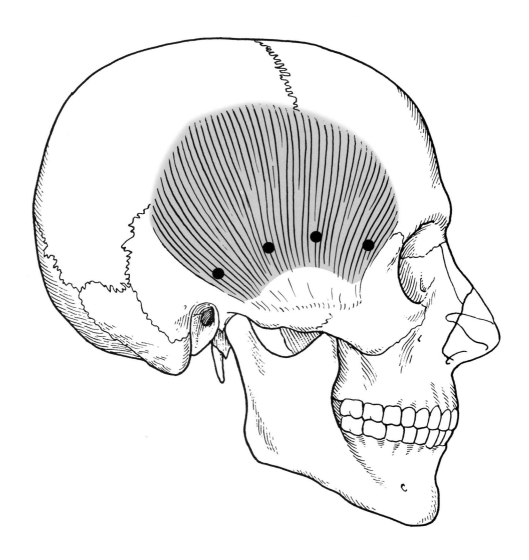

Origin: Temporal fossa and temporal fascia

Insertion: Coronoid process of mandible via a tendon that passes deep to the zygomatic arch

Action: Elevates and retracts mandible, assists in side to side movement of mandible

Innervation: Mandibular branch of trigeminal nerve (V)

The **temporalis** is a fan-shaped muscle that covers parts of the temporal, frontal and parietal bone. It maintains the jaw position at rest. The **trigger points** for this muscle are located anteriorly, medially and posteriorly along the inferior aspect of the muscle near the tendon at the coronoid process of the mandible. The **referred pain patterns** are the temporal region, eyebrow and upper teeth.

Masseter

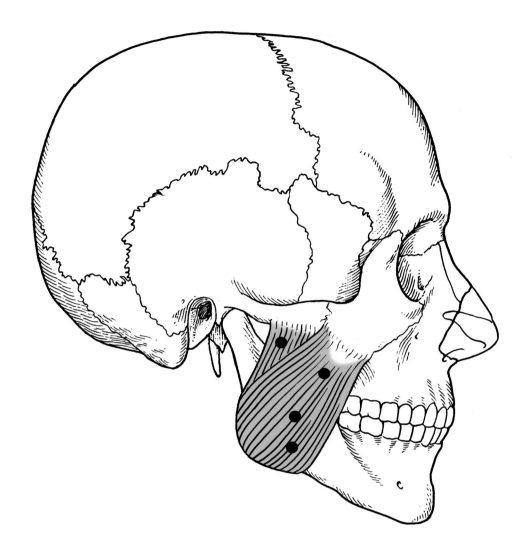

Origin: Zygomatic process of maxilla and medial and inferior surfaces of zygomatic arch

Insertion: Angle and ramus of the mandible

Action: Elevates mandible and slightly protracts it

Innervation: Mandibular branch of trigeminal nerve (V)

This cheek muscle bulges as it elevates the lower jaw when you clench your teeth and close the teeth when chewing. The **trigger points** for this muscle are at the tendinous junction near the zygomatic arch and in the belly of the muscle. The **referred pain patterns** are the upper jaw, the ear, and the eyebrow.

Pterygoideus Medialis

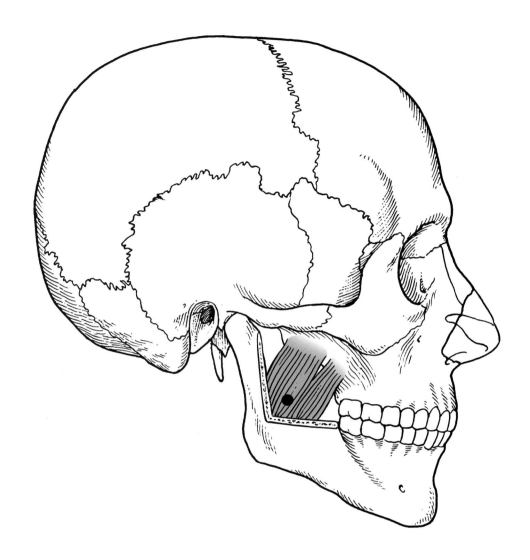

Origin: Medial surface of lateral pterygoid plate of sphenoid, and the maxilla and palatine bones

Insertion: Posterioinferior aspect of the medial surface of ramus and angle of the mandible

Action: Synergistic with temporalis and masseter in elevation of mandible. It causes protrusion and side to side movements of the mandible.

Innervation: Mandibular branch of trigeminal nerve (V)

This is a deep, two-headed muscle that runs along the inner surface of the mandible and is largely concealed by that bone. Together with the **pteryogoideus lateralis**, this muscle is primarily responsible for the side-to-side motions involved in chewing and grinding the teeth. The **trigger point** for this muscle is in the belly of the muscle, and it is best accessed from inside the mouth. The **referred pain pattern** is the back of the throat, into the ear.

Pterygoideus Lateralis

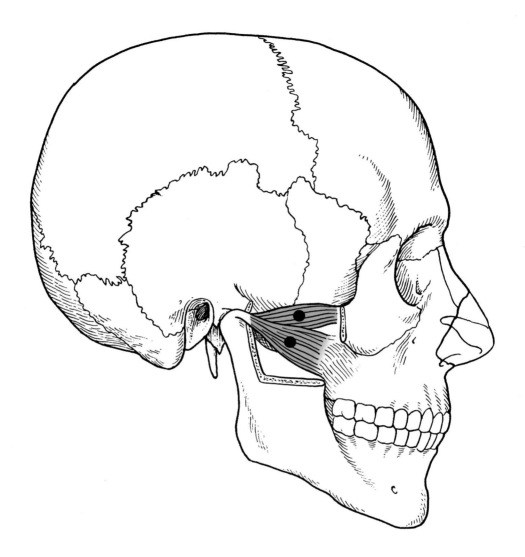

Origin: A superior head arises from the greater wing of sphenoid bone; an inferior head from the lateral surface of the lateral pterygoid plate of sphenoid

Insertion: Both heads insert on the Mandibular condyle and temporomandibular joint capsule

Action: Protrudes, depresses and moves mandible from side-to-side

Innervation: Mandibular branch of trigeminal nerve (V)

The **trigger points** for this muscle are the bellies of both divisions; the **referred pain pattern**s are the cheek and the temporomandibular joint. Constant **trigger point** generated tension tends to pull the lower jaw forward and disarticulate the joint.

Extrinsic Tongue (or Glossal) and Pharyngeal Muscles

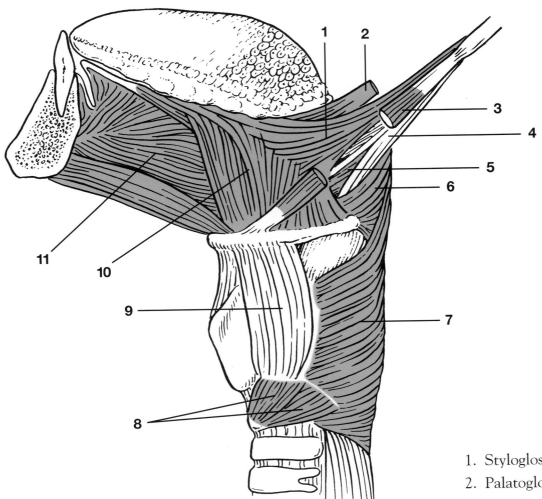

1. Styloglossus
2. Palatoglossus
3. Stylohyoid
4. Stylopharyngeus
5. Superior pharyngeal constrictor
6. Middle pharyngeal constrictor
7. Inferior pharyngeal constrictor
8. Cricothyroid
9. Thyrohyoid
10. Hyoglossus
11. Genioglossus

Intrinsic tongue muscles (those totally within the tongue) change the shape of the tongue, important in chewing and speaking, but do not move it. Extrinsic tongue muscles arise from bony structures and project up into the tongue to move it anteriorly, posteriorly and laterally.

The muscle cells of the inner muscle layer of the pharynx run superior-inferior. The three circular muscles encircle the inner layer and are arranged one above another, each slightly overlapping the one above. Sequential contraction of these three muscles from superior to inferior produces the force that moves food through the pharynx into the esophagus below.

An Illustrated Atlas of the Skeletal Muscles

Extrinsic Tongue (or Glossal) and Pharyngeal Muscles

Styloglossus

Origin: Styloid process of temporal bone

Insertion: Lower lateral portion of tongue

Action: Retracts and elevates tongue

Innervation: Hypoglossal cranial nerve (XII)

Stylopharyngeus

Origin: Medial side of base of styloid process

Insertion: Lateral aspects of pharynx and thyroid cartilage

Action: Elevates larynx and dilates pharynx to help bolus descend

Innervation: Glossopharyngeal nerve (IX)

Superior Pharyngeal Constrictor

Origin: Pterygoid hamulus, pterygomandibular raphe, posterior end of mylohyoid, and side of tongue

Insertion: Median raphe of pharynx and pharyngeal tubercle

Action: Constrict wall of pharynx during swallowing

Innervation: Pharyngeal and superior laryngeal branches of vagus nerve (X)

Middle Pharyngeal Constrictor

Origin: Stylohyoid ligament and greater and lesser horns of hyoid bone

Insertion: Median raphe of pharynx

Action: Constricts wall of pharynx during swallowing

Innervation: Same as above

Inferior Pharyngeal Constrictor

Origin: Oblique line of thyroid cartilage and side of cricoid cartilage

Insertion: Median raphe of pharynx

Action: Constricts wall of pharynx during swallowing

Innervation: Same as above

Genioglossus

Origin: Internal surface of mandible near symphysis

Insertion: Lower portion of tongue and body of hyoid bone

Action: Protracts tongue; can also depress tongue and work with other extrinsic muscles to retract tongue

Innervation: Hypoglossal cranial nerve (XII)

Hyoglossus

Origin: Body and greater horn of hyoid bone

Insertion: Lower lateral portion of tongue

Action: Depresses tongue and draws its side downward

Innervation: Hypoglossal cranial nerve (XII)

Tympanic Cavity Muscles

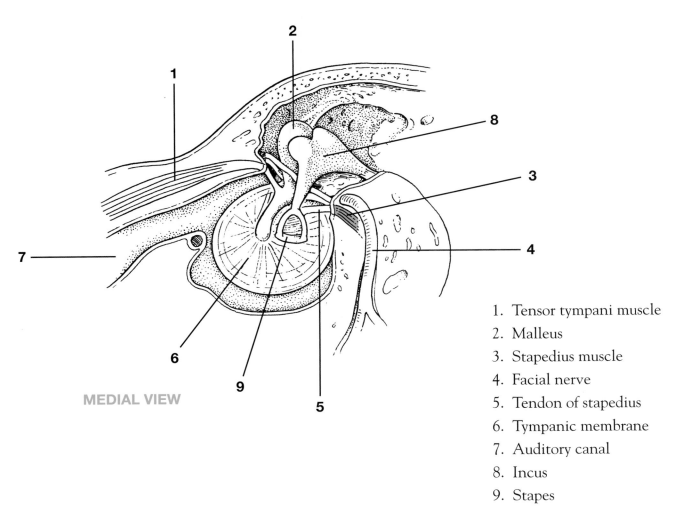

MEDIAL VIEW

1. Tensor tympani muscle
2. Malleus
3. Stapedius muscle
4. Facial nerve
5. Tendon of stapedius
6. Tympanic membrane
7. Auditory canal
8. Incus
9. Stapes

Tensor Tympani

Origin:	Wall of auditory tube
Insertion:	On the malleus ossicle
Action:	(See below)
Innervation:	Trigeminal nerve (V)

Stapedius

Origin:	Posterior wall of middle ear cavity
Insertion:	Onto the stapes ossicle
Action:	(See below)
Innervation:	Facial cranial nerve (VII)

The three middle ear ossicles (**malleus, incus, stapes**) articulate with one another by mini synovial joints and span from the tympanic membrane to the bony enclosure of the inner ear. They are suspended by tiny ligaments, and their movement transmits the vibrations of the eardrum to the oval window, which in turn sets fluids of the inner ear into motion, ultimately stimulating the hearing or auditory receptors. The **tensor tympani** and **stapedius** muscles contract reflexively in response to very loud sounds, preventing damage to the hearing receptors. The tensor tympani tenses the eardrum by pulling it medially; the stapedius dampens excessive vibrations to the whole chain of three bones, thus limiting the movement of the stapes in the oval window.

Laryngeal Muscles

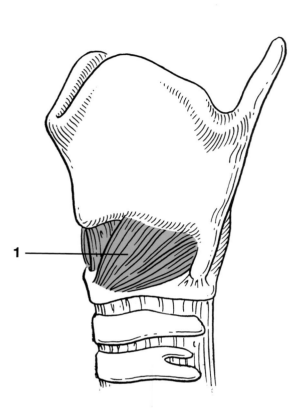

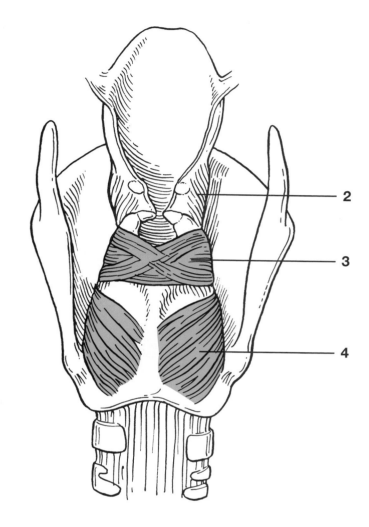

1. Cricothyroideus
2. Thyroarytenoideus
3. Arytenoid
4. Cricoarytenoideus

Arytenoid

Origin: Arytenoid cartilage on one side

Insertion: Arytenoid cartilage on opposite side

Action: Close laryngeal aditus by approximating arytenoid cartilages

Innervation: Laryngeal branch of vagus nerve (X)

Cricoarytenoideus

Origin: Arch of cricoid cartilage

Insertion: Muscular process of arytenoid cartilage

Action: Adducts vocal folds

Innervation: Laryngeal branch of vagus nerve (X)

Cricothyroideus

Origin: Anterolateral part of cricoid cartilage

Insertion: Inferior margin and inferior horn of thyroid cartilage

Action: Stretches and tenses the vocal folds

Innervation: Laryngeal branch of vagus nerve (X)

Thyroarytenoideus

Origin: Posterior surface of thyroid cartilage

Insertion: Muscular process of arytenoid process

Action: Relaxes vocal folds

Innervation: Laryngeal branch of vagus nerve (X)

Rectus Extrinsic Eye Muscles

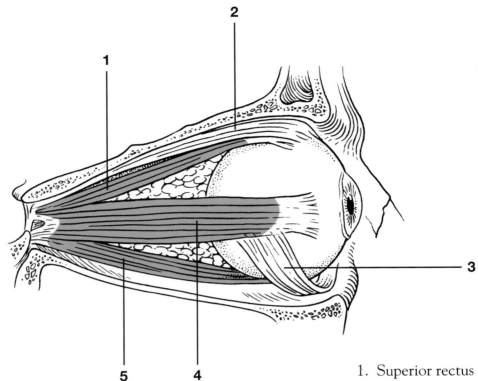

1. Superior rectus
2. Levator palpebrae superioris
3. Inferior oblique
4. Lateral rectus
5. Inferior rectus muscle

Superior Rectus

Origin: Annular ring at the back of the bony orbit

Insertion: Superior surface of anterior sclera

Action: Elevates eyeball

Innervation: Oculomotor nerve (III)

Inferior Rectus

Origin: Annular ring at the back of the bony orbit

Insertion: Inferior surface of anterior sclera

Action: Depresses eyeball

Innervation: Oculomotor nerve (III)

Medial Rectus

Origin: Annular ring at the back of the bony orbit

Insertion: Medial surface of the anterior sclera

Action: Medially rotates eyeball

Innervation: Oculomotor nerve (III)

Lateral Rectus

Origin: Annular ring at the back of the bony orbit

Insertion: Lateral surface of the anterior sclera

Action: Laterally rotates eyeball

Innervation: Abducens nerve (VI)

Oblique Extrinsic Eye Muscles

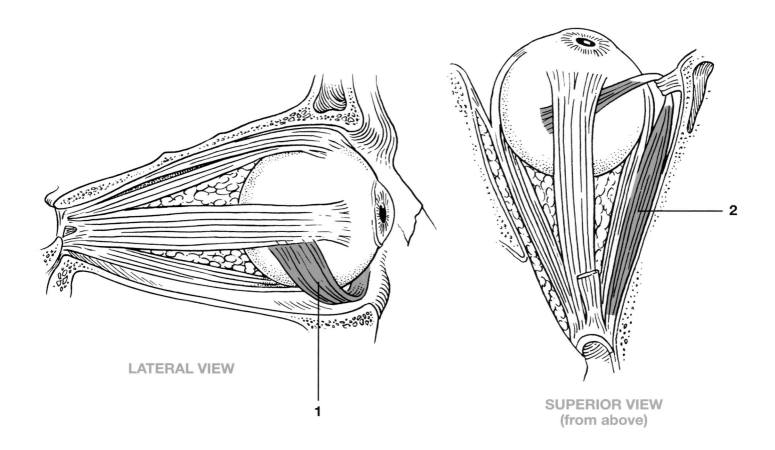

LATERAL VIEW

SUPERIOR VIEW
(from above)

1. Inferior oblique
2. Superior oblique

Superior Oblique Muscle

Origin: Tendinous ring attached to the bony orbit around the optic foramen

Insertion: Through a fibrocartilaginous ring, the trochlea, and attaches on the superior surface of the eyeball between the superior and lateral rectus muscles

Action: Rotates the eyeball moving it downward and laterally

Innervation: Trochlear nerve (IV)

Inferior Oblique Muscle

Origin: Maxillary bone at the medial inferior corner of the orbit

Insertion: Lateral surface of the eyeball between the inferior and lateral rectus muscles

Action: Rotates the eyeball moving it upward and laterally

Innervation: Oculomotor nerve (III)

Muscles of the Soft Palate

1. Tensor veli palatini
2. Levator veli palatini
3. Palatopharyngeus
4. Musculus uvulae
5. Salpingopharyngeus
6. Tonsils

Tensor veli palatini

Origin:	Scaphoid fossa of medial pterygoid plate, spine of sphenoid bone and cartilage of auditory tube
Insertion:	Palatine aponeurosis
Action:	Tenses soft palate and opens mouth of auditory tube during swallowing and yawning
Innervation:	Medial pterygoid branch of the mandibular branch of trigeminal nerve (V)

Levator veli palatini

Origin:	Cartilage of auditory tube and petrous part of temporal bone
Insertion:	Palatine aponeurosis
Action:	Elevates soft palate during swallowing and yawning
Innervation:	Pharyngeal branch of vagus nerve (X)

Palatopharyngeus

Origin:	Hard palate and palatine aponeurosis
Insertion:	Lateral wall of pharynx
Action:	Tenses soft palate and pulls walls of pharynx superiorly, anteriorly and medially during swallowing
Innervation:	Cranial part of spinoaccessory (XI) nerve and pharyngeal branch of vagus nerve (X)

Musculus uvulae

Origin:	Posterior nasal spine and palantine aponeurosis
Insertion:	Mucosa of uvula
Action:	Shortens uvula and pulls it superiorly
Innervation:	Cranial part of spinoaccessory nerve (XI) and pharyngeal branch of vagus nerve (X)

Salpingopharyngeus

Origin:	Cartilage of auditory tube
Insertion:	Posterior border of thyroid cartilage and side of esophagus
Action:	Elevates larynx and pharynx during swallowing and speaking
Innervation:	Glossopharyngeal nerve (IX)

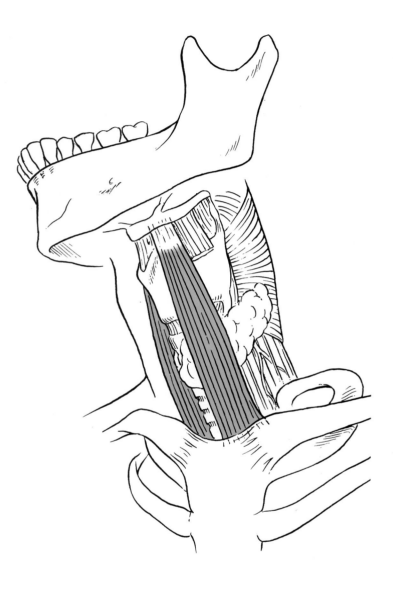

Muscles of the Neck

Platysma

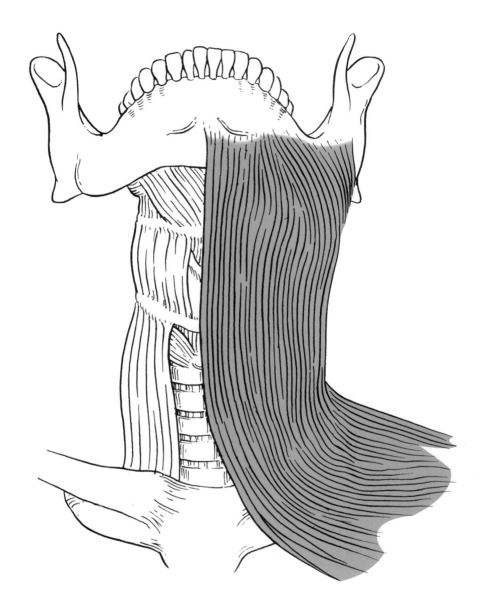

Origin: Subcutaneous fascia covering the pectoralis major and the deltoideus

Insertion: Lower margin of mandible, and subcutaneous fascia and muscles of jaw and mouth

Action: Draws down the lower lip and angle of mouth, tenses skin of neck; helps depress mandible

Innervation: Cervical branch of facial nerve (VII)

This is the muscle that tenses the neck when shaving.

Sternocleidomastoideus

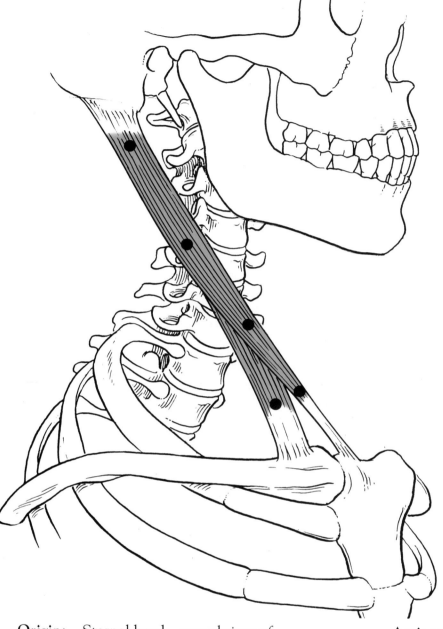

The fleshy parts of this strap-shaped muscle divides the cervical region into anterior and posterior triangles. It is a key muscular landmark in the neck. Spasms in one of these muscles may cause "wryneck" or torticollis. There are several **trigger points** along the entire length of both heads of the muscle. The **referred pain pattern** is head and face especially in the occipital region, ear and forehead.

Origin: Sternal head—manubrium of sternum

Clavicular head—superior border of medial third of clavicle

Insertion: Mastoid process of temporal and lateral half of superior nuchal line

Action: Contraction of one side—bends neck laterally and rotates head to opposite side

Contraction of both sides together —flexes neck; with head fixed it assists in elevating the thorax during forced inspiration

Innervation: Spinal part of spinoaccessory nerve (XI) and branches of cervical spinal nerves (C2–C4)

Digastricus

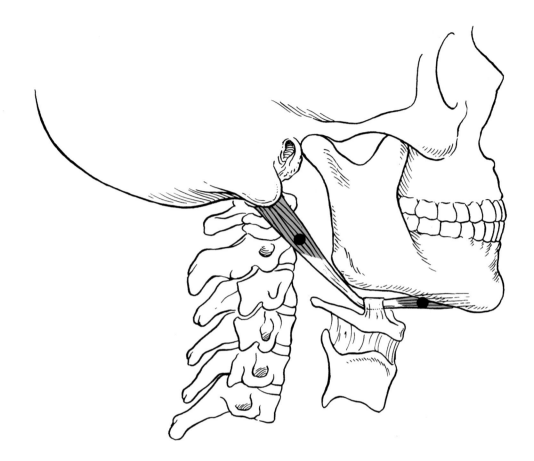

Origin: Posterior belly—mastoid notch of temporal bone

Anterior belly—inner side of inferior margin of mandible near mandibular symphysis

Insertion: Both bellies insert on the body of the greater cornu of the hyoid bone by a fibrous loop

Action: Acting together, the digastric muscles elevate the hyoid bone and steady it during swallowing and speech. The posterior belly helps open the mouth and depresses the mandible

Innervation: Anterior belly—mandibular branch of trigeminal nerve (V)

Posterior belly—cervical branch of facial nerve (VII)

The **trigger points** for this muscle are in the belly of each division of the muscle. The **referred pain pattern** is in the sternocleidomastoid area and the bottom front teeth.

Stylohyoideus

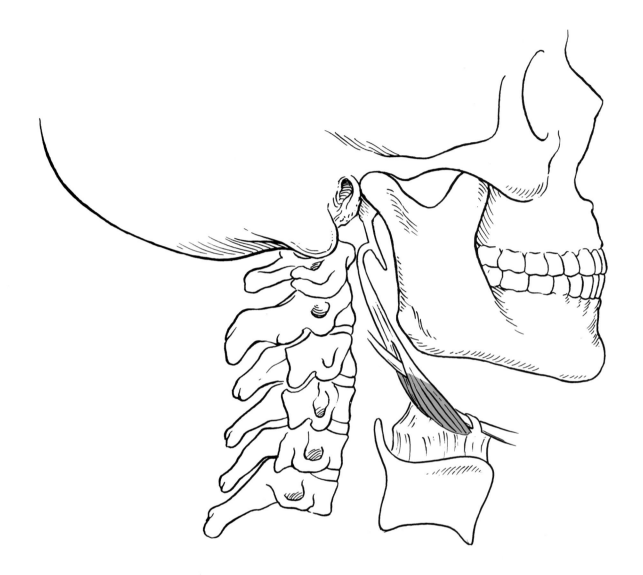

Origin: Styloid process of temporal bone

Insertion: Body of hyoid bone at junction of greater cornu

Action: Elevates and retracts the hyoid elongating the floor of the mouth and lifts the tongue during swallowing

Innervation: facial nerve (VII)

Mylohyoideus

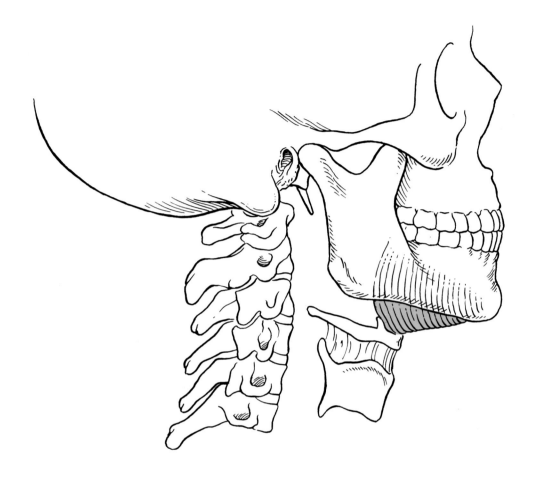

Origin: Mylohyoid line of mandible

Insertion: Upper border and median raphe of hyoid bone

Action: Elevates hyoid bone and raises floor of mouth and tongue

Innervation: Mandibular branch of trigeminal nerve (V)

Geniohyoideus

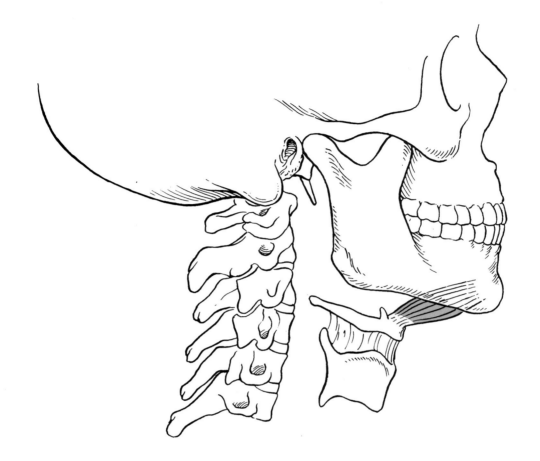

Origin: Inferior mental spine on inner surface of mandible

Insertion: Anterior surface of body of hyoid bone

Action: Pulls hyoid bone superiorly and anteriorly shortening the floor of the mouth. It draws the tongue forward.

Innervation: First cervical nerve (C1) through the hypoglossal nerve

Sternohyoideus

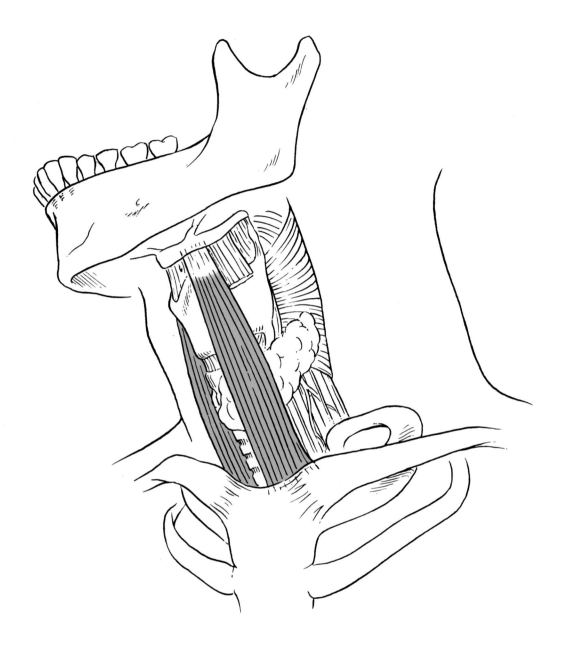

Origin: Medial end of clavicle and manubrium of sternum

Insertion: Lower margin of the body of the hyoid bone

Action: Depresses hyoid bone if it has been elevated, as in swallowing

Innervation: Cervical spinal nerve C1–C3 through the ansa cervicalis (slender nerve root in cervical plexus)

Sternothyroideus

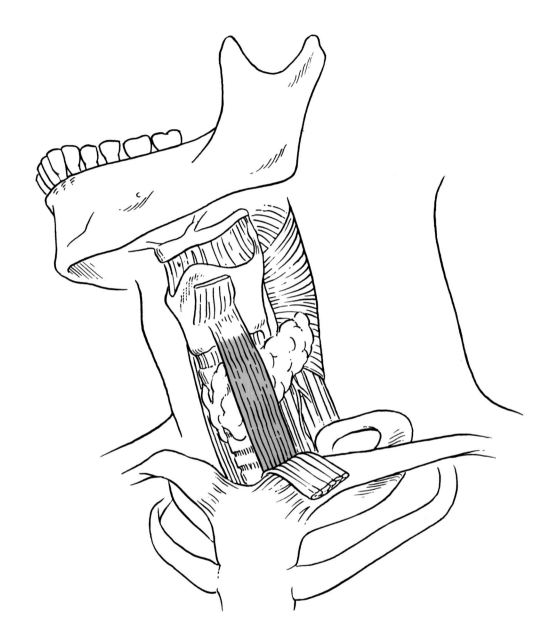

Origin: Posterior surface of manubrium of sternum

Insertion: Oblique line on lamina of thyroid cartilage

Action: Depresses larynx

Innervation: Ansa cervicalis (C1–C3)

Thyrohyoideus

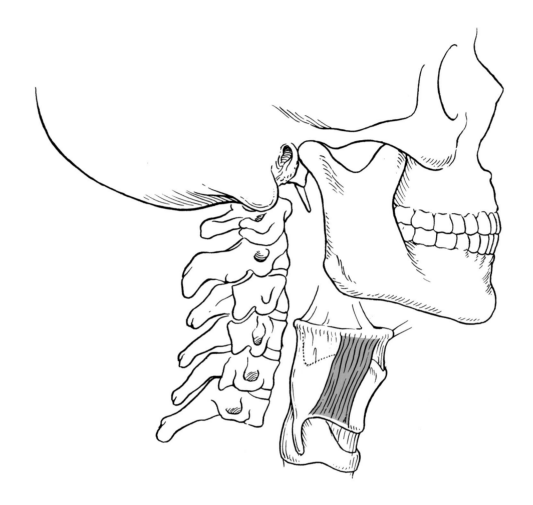

Origin: Lamina of the thyroid cartilage at the oblique line

Insertion: Greater cornu of hyoid bone

Action: Depresses hyoid and elevates larynx if hyoid is fixed.

Innervation: First cervical nerve through the hypoglossal nerve (XII)

Omohyoideus

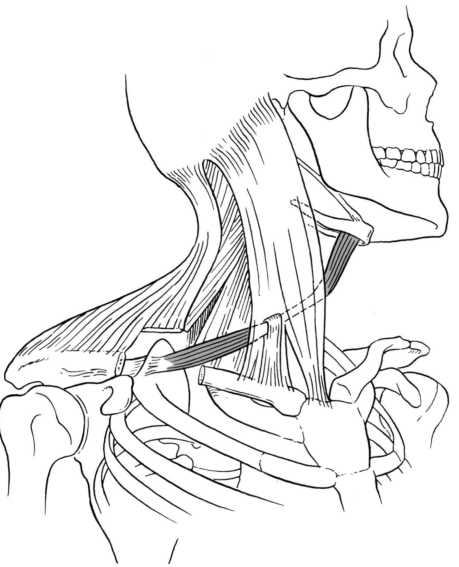

Origin: Inferior belly—superior border of scapula near scapula notch and suprascapula ligaments

Superior belly—arises from tendon of inferior belly near sternocleidomastoid

Insertion: Inferior belly—ends as a tendon (bound to clavicle by central tendon)

Superior belly—inserts on lower border of hyoid bone

Action: Depresses and retracts hyoid bone; retracts larynx

Innervation: Ansa cervicalis (C2, C3)

Like the digastricus muscle, the **omohyoid** is a straplike muscle with two bellies united by an intermediate tendon. It is lateral to the sternohyoid.

Longus Capitis

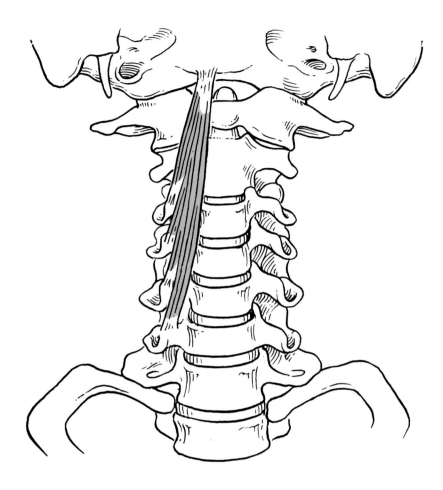

Origin: Anterior tubercles of the transverse processes of the 3rd through 6th cervical vertebrae

Insertion: Basilar process of occipital bone anterior to foramen magnum

Action: Flexes cervical vertebrae and head

Innervation: C1–C4

Longus Coli

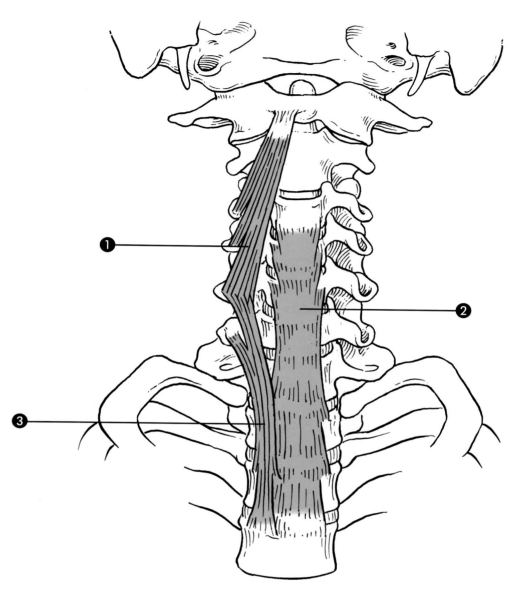

1. Superior Oblique Part

Origin: Transverse processes of the third, fourth, and fifth cervical vertebrae

Insertion: Anterior arch of Atlas

2. Medial Part

Origin: Anterior surfaces of the bodies of the first three thoracic and lower three cervical vertebrae

Insertion: Anterior surfaces of the second, third and fourth cervical vertebrae

Action: All three parts flex the neck. The oblique portion bends it laterally. Inferior oblique portion rotates it to the opposite side.

Innervation: C2–C7

3. Inferior Oblique Part

Origin: Anterior surface of the bodies of the first two or three thoracic vertebrae

Insertion: Transverse processes of the fifth and sixth cervical vertebrae

---

Rectus Capitis Anterior

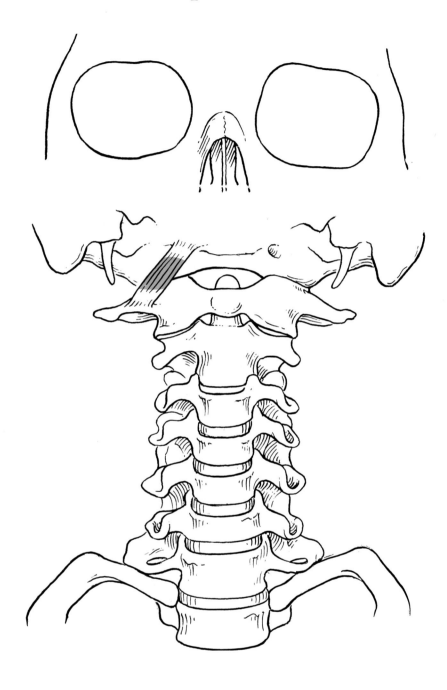

Origin: Anterior base of transverse process of atlas

Insertion: Occipital bone anterior to the foramen magnum

Action: Flexes head

Innervation: C1–C2

Rectus Capitis Lateralis

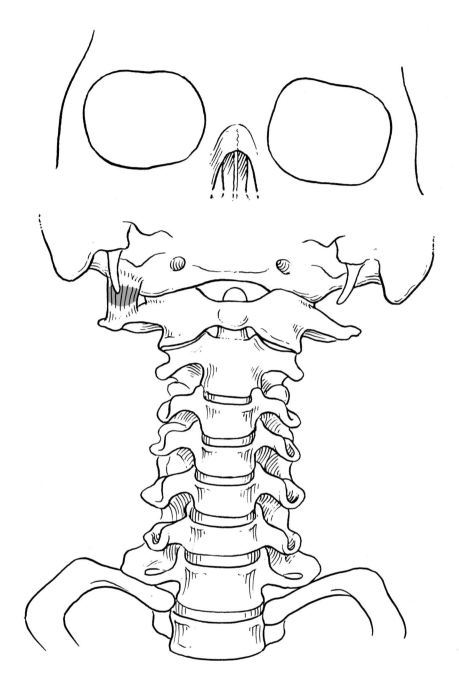

Origin: Transverse process of atlas

Insertion: Jugular process of occipital bone

Action: Bends head laterally

Innervation: C1, C2

Rectus Capitis Posterior Major

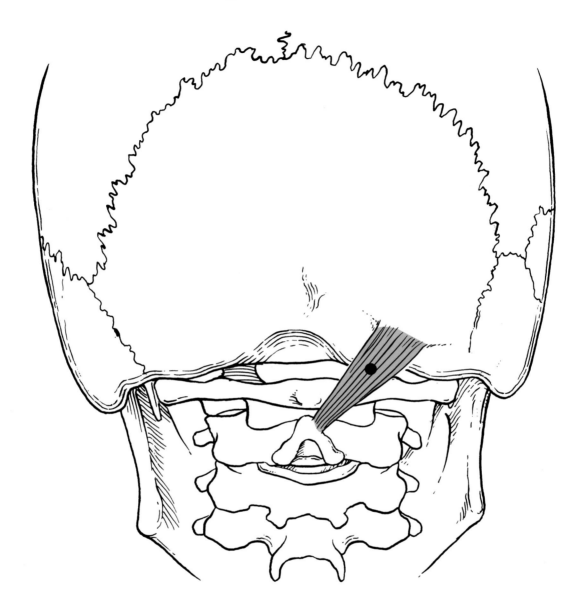

Origin: Spinous process of axis

Insertion: Lateral portion of inferior nuchal line of occipital bone

Action: Extends and rotates the head toward the same side

Innervation: Dorsal ramus of the first (suboccipital) cervical nerve (C1)

The spasmodic contraction of four of the small cervical muscles set off the **referred pain pattern** commonly associated with a **tension headache.** These muscles are the **rectus capitis posterior major, rectus capitis posterior minor, obliquus capitis superior** and **obliquus capitis inferior.** These muscles are especially important in the movement of the upper two cervical vertebrae. The **trigger point** is in the belly of the muscle.

Rectus Capitis Posterior Minor

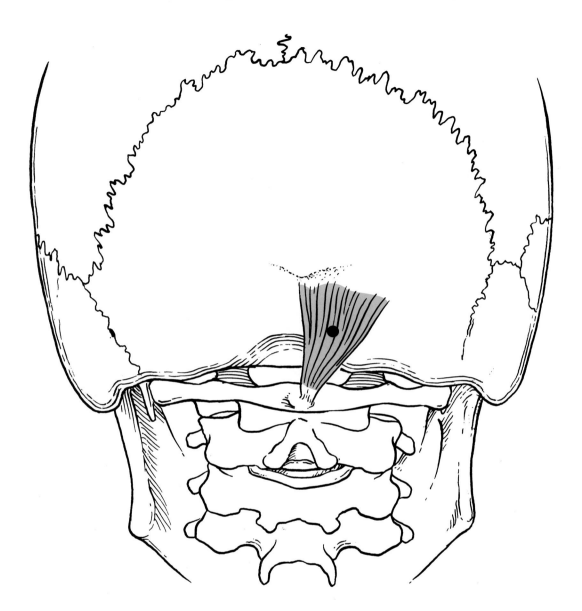

Origin: Posterior tubercle of atlas

Insertion: Median portion of inferior nuchal line of occipital bone

Action: Extends the head?

Innervation: Dorsal ramus of first cervical (suboccipital) nerve (C1)

There is difference of opinion about the action of this muscle. It is generally assumed that it extends the head, but **myographic studies** indicate that it does not act in extension, but rather functions as a restraint to flexion and forward movement of the head. The **trigger point** is in the belly of the muscle.

Scalenus Anterior

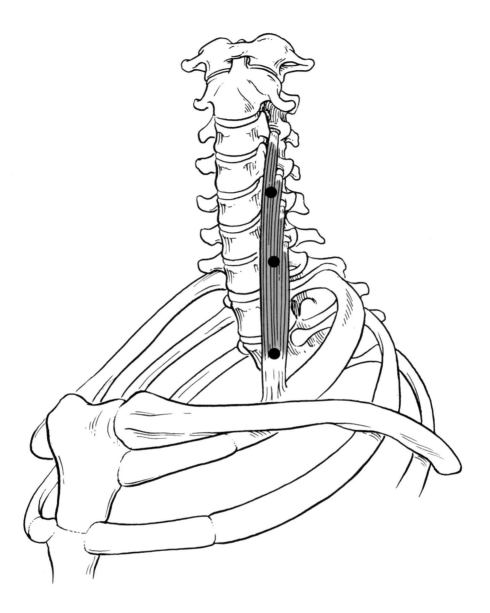

Origin: Anterior tubercle of the transverse processes of the third through sixth cervical vertebrae

Insertion: Scalene tubercle on the inner border and upper surface of the first rib

Action: Bends the cervical portion of the vertebral column forward and laterally. It also assists in the elevation of the first rib.

Innervation: Ventral rami of the fourth through sixth cervical nerves (C4–C6)

There are multiple **trigger points** along the length of the muscle. The **referred pain pattern** is along the upper and lower arm, lateral side of the hand and just lateral to the midline in both the anterior and posterior upper thorax.

An Illustrated Atlas of the Skeletal Muscles

Scalenus Medius

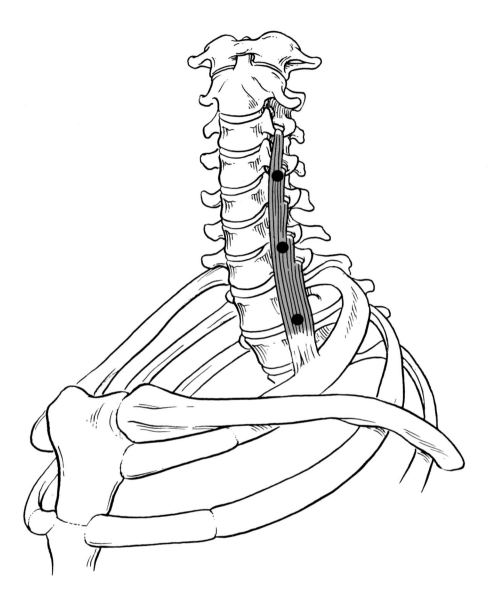

Origin: Front of the posterior tubercles of the transverse processes of the second through seventh cervical vertebrae

Insertion: Upper surface of the first rib

Action: Acting from above, it helps to raise the first rib. Acting from below, it laterally flexes the neck.

Innervation: Ventral rami of the fourth through the eighth cervical nerve (C4–C8)

There are multiple **trigger points** along the length of the muscle. The **referred pain pattern** is along the upper and lower arm, lateral side of the hand and just lateral to the midline in both the anterior and posterior upper thorax.

Scalenus Posterior

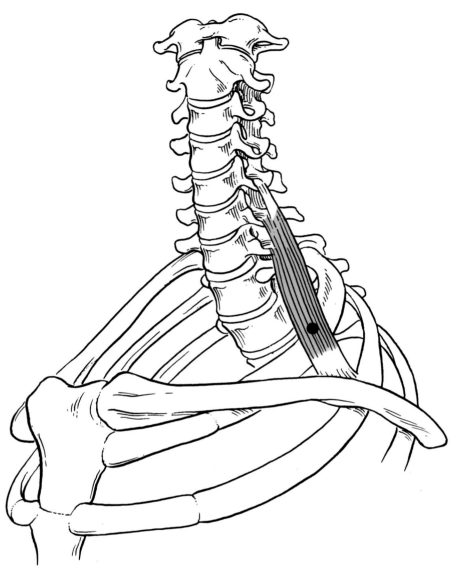

Origin: Posterior tubercles of the transverse processes of the fourth through sixth cervical vertebrae

Insertion: Outer surface of second rib

Action: Raises second rib. When the second rib is fixed, bends the lower end of the cervical portion of the vertebral column to the same side

Innervation: Ventral rami of the sixth through eighth cervical nerves (C6–C8)

The **scalenus group** acts as an accessory muscle of respiration. By raising the first and second ribs, they assist in inspiration. The **trigger point** for each of these muscles is in the belly near the rib attachment points. The **referred pain pattern** is in the pectoral region, the rhomboid region and the entire length of the arm to the hand.

Obliquus Capitis Inferior

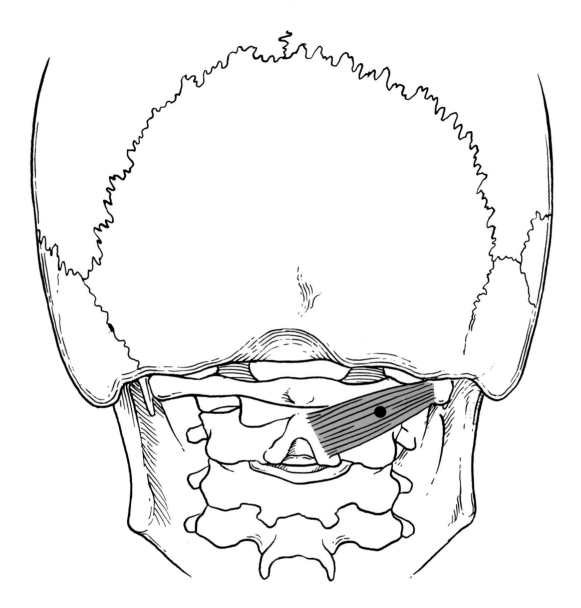

Origin: Spinous process and upper lamina of axis

Insertion: Transverse process of the atlas

Action: Rotates head

Innervation: Dorsal ramus of the first cervical (suboccipital) nerve (C1)

The **trigger point** is in the belly of the muscle.

Obliquus Capitis Superior

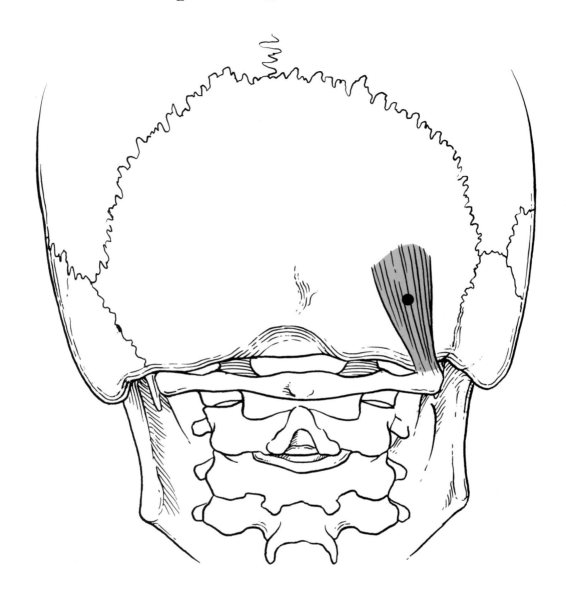

Origin: Superior surface of transverse process of atlas

Insertion: Occipital bone between inferior and superior nuchal lines

Action: Bends the head backward and laterally to the same side

Innervation: Dorsal ramus of the first cervical (suboccipital) nerve (C1)

The **trigger point** is in the belly of the muscle.

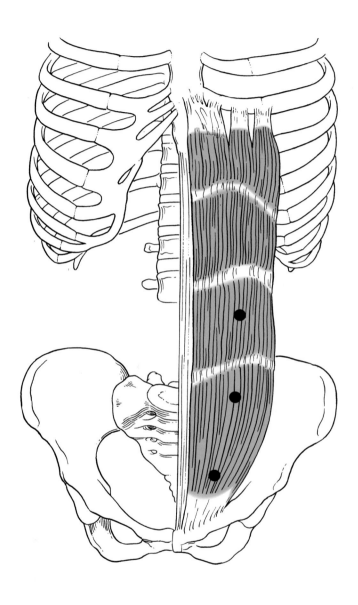

6

Muscles of the Torso

Splenius Capitis

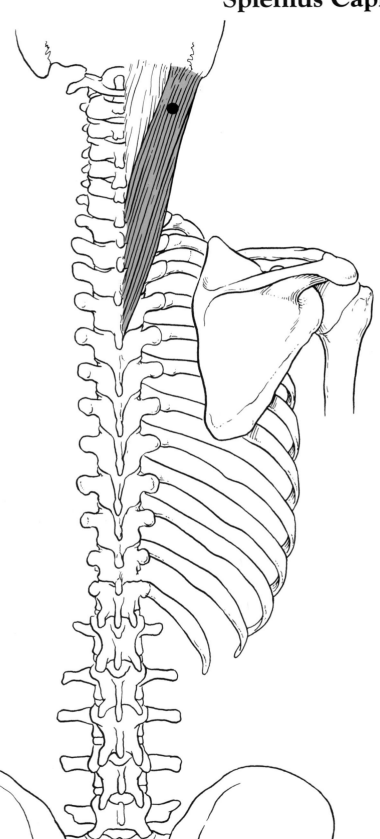

Origin: Fascia and spinous processes of seventh cervical and first four thoracic vertebrae (C7–T4)

Insertion: Lateral one third of the superior nuchal line and the mastoid process of the temporal bone

Action: Extends and hyperextends the head. Contraction of one side only laterally flexes and rotates the head and neck.

Innervation: Dorsal rami of the middle cervical nerves (C4–C8)

The word **splenius** means bandage. The splenius muscles seem to wrap around the deeper neck muscles. The **trigger point** for the muscle is in the belly close to the head. The **referred pain pattern** is to the top of the head and eye region. The muscle can be **palpated** with difficulty on the neck between the trapezius and sternocleidomastoid muscles.

Splenius Cervicis

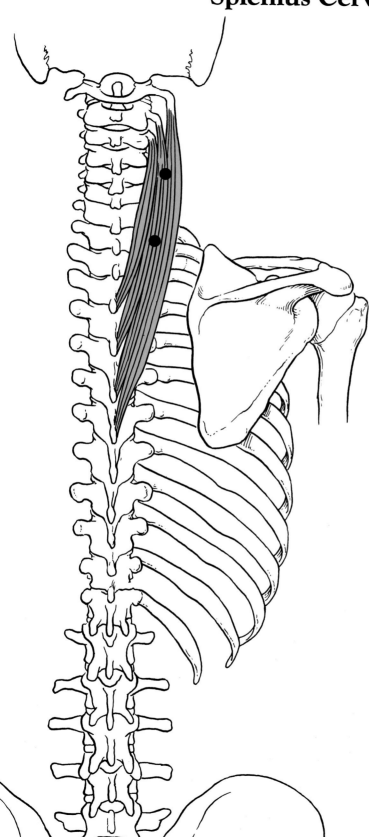

Origin: Spinous processes of third through sixth thoracic vertebrae (T3–T6)

Insertion: Posterior tubercles of the transverse processes of the first three cervical vertebrae (C1–C3)

Action: The muscles extend and hyperextend the neck. Contraction of one side only laterally flexes and rotates the neck and head.

Innervation: Dorsal rami of the lower cervical nerves (C4–C8)

The **trigger points** are in the belly of the muscle and near the insertion.

Erector Spinae

1. Iliocostalis cervicis

Origin: Angles of the third through sixth ribs

Insertion: Posterior tubercles of the transverse processes of the fourth through sixth cervical vertebrae (C4–C6)

Action: Extension, lateral flexion of the vertebral column

Innervation: Dorsal rami of the lower cervical and thoracic spinal nerves

2. Iliocostalis thoracis

Origin: Angles of lower six ribs medial to the iliocostalis lumborum

Insertion: Superior border at the angles of the upper six ribs

Action: Extension, lateral flexion of vertebral column

Innervation: Dorsal rami of the thoracic spinal nerves

3. Iliocostalis lumborum

Origin: Medial and lateral sacral crests and medial part of iliac crest

Insertion: Angles of lower six ribs

Action: Extension, lateral flexion of vertebral column, lateral movement of pelvis

Innervation: Dorsal rami of the thoracic and lumbar spinal nerves

The **erector spinae** muscles are a group of three sets of muscles: the **iliocostalis, longissimus and spinalis.** Together they extend and laterally flex the vertebral column. In the lumbar region they lie deep to the **lumbodorsal fascia** (thoracolumbar) and in the thoracic region they are deep to the **trapezius** and **rhomboideus** muscles. As a group they can easily be **palpated** along the entire length of the vertebral column.

Erector Spinae

1. Longissimus cervicis

Origin: Transverse processes of upper five thoracic vertebrae (T1–T5)

Insertion: Posterior tubercles of the transverse processes of the second through sixth cervical vertebrae (C2–C6)

Action: Extension and lateral flexion of vertebral column

Innervation: Dorsal rami of thoracic spinal nerves

2. Longissimus capitis

Origin: Transverse processes of upper five thoracic vertebrae (T1–T5) and articular processes of lower four cervical vertebrae

Insertion: Posterior aspect of mastoid process of temporal bone

Action: Extends and rotates the head

Innervation: Dorsal rami of middle and lower cervical nerves

3. Longissimus thoracis

Origin: Aponeurosis and transverse processes of lumbar and lower thoracic vertebrae

Insertion: Transverse processes of all thoracic vertebrae and between tubercles and angles of lower ten ribs

Action: Extension and lateral flexion of vertebral column

Innervation: Dorsal rami of thoracic and lumbar spinal nerves

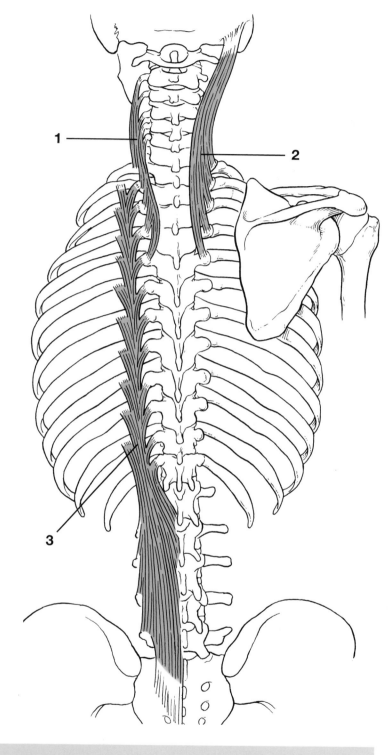

During full flexion, when bending over, the **erector spinae** muscles are relaxed. Upon standing upright, these muscles are initially inactive and extension is initiated by the hamstring muscles. As a result of this, lifting a load from the bent over position can cause injury to these muscles.

Erector Spinae

1. *Spinalis capitis*

Origin: Transverse processes of upper seven thoracic (T1–T7) and articular processes of fourth through seventh cervical vertebrae (C4–C7)

Insertion: Between superior and inferior nuchal lines of the occipital bone

Action: Extends the vertebral column

Innervation: Dorsal rami of lower cervical and thoracic spinal nerves

2. *Spinalis cervicis*

Origin: Spinous process of first and second thoracic (T1, T2) and seventh cervical vertebrae (C7)

Insertion: Spinous processes of second and third cervical vertebrae (C2, C3)

Action: Extends the vertebral column

Innervation: Dorsal rami of lower cervical and thoracic spinal nerves

3. *Spinalis thoracis*

Origin: Spinous processes of the lower two thoracic and upper two lumbar vertebrae

Insertion: Spinous process of upper eight thoracic vertebrae (T1–T8)

Action: Extends the vertebral column

Innervation: Dorsal rami of thoracic and lumbar spinal nerves

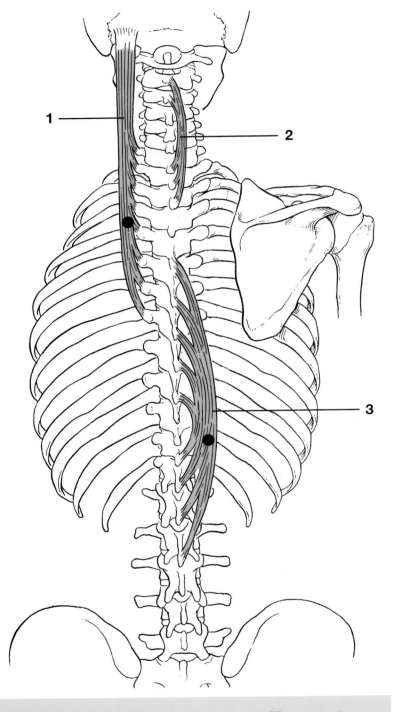

The **erector spinae** muscles go into powerful spasms following injury to back structures. **Trigger points** for this group are usually found in the midscapular and lumbar regions. The **referred pain pattern** is to the scapular, lumbar, gluteal and abdominal regions. To help **palpate** the erector spinae group, ask the individual being examined to arch their back slightly or raise their feet slightly.

An Illustrated Atlas of the Skeletal Muscles

Transversospinalis

1. Semispinalis capitis

Origin: Transverse process of the upper six thoracic (T1–T6) and seventh cervical (C7) and articular processes of the fourth through six cervical vertebrae (C4–C6)

Insertion: Between the superior and inferior nuchal lines of the occipital bone

Action: Extension of the head and rotation to the opposite side

Innervation: Dorsal rami of the first six cervical spinal nerves (C1–C6)

2. Semispinalis cervicis

Origin: Transverse processes of upper six thoracic (T1–T6) and articular processes of lower four cervical vertebrae

Insertion: Spinous process of second through fifth cervical vertebrae (C2–C5)

Action: Extension and rotation of vertebral column

Innervation: Dorsal rami of lower three cervical spinal nerves

3. Semispinalis thoracis

Origin: Transverse processes of lower six thoracic vertebrae

Insertion: Spinous processes of the lower two cervical and upper four thoracic vertebrae (T1–T4)

Action: Extension and rotation of vertebral column

Innervation: Dorsal rami of upper six thoracic spinal nerves (T1–T6)

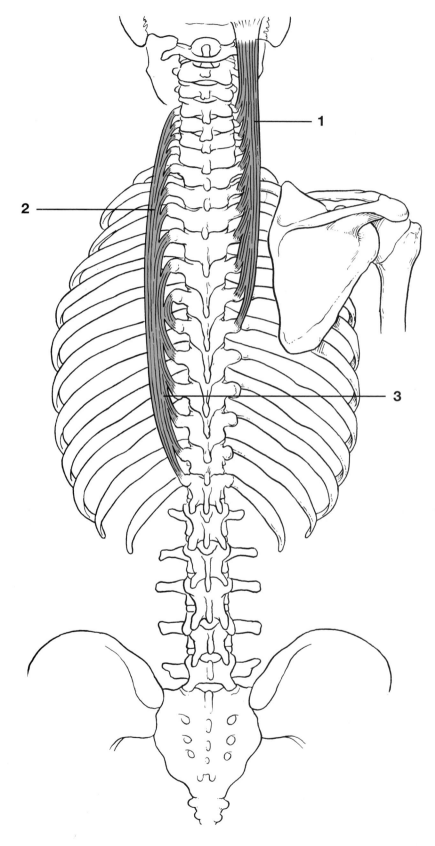

Multifidis

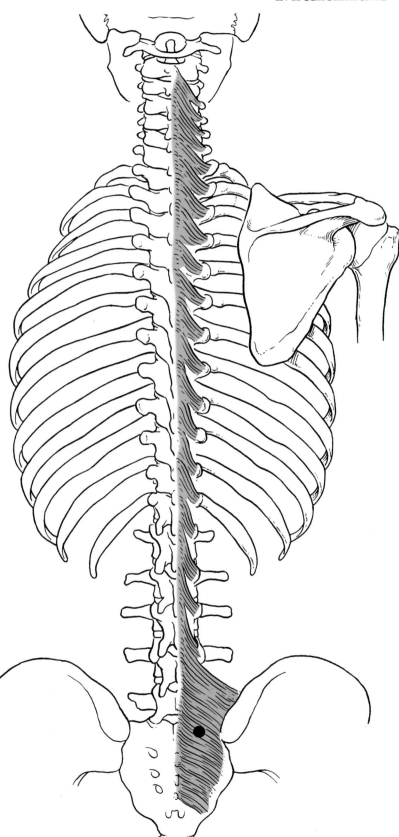

Origin: Articular processes of the last four cervical, transverse processes of all thoracic, and mammillary processes of lumbar vertebrae, the posterior superior iliac spine, posterior sacroiliac ligaments and dorsal surface of sacrum adjacent to sacral spinous processes

Insertion: Spinous process of the vertebra above the vertebra of origin

Action: Extend and rotate vertebral column

Innervation: Dorsal rami of spinal nerves

The **multifidi** are part of the **transversospinalis** group of muscles. They lie deep to the **erector spinae**. This group extends and rotates the spine. The **trigger point** is in the belly of the muscle in the lumbosacral region. The **referred pain pattern** may feel like it is in the spine itself because tension in these small diagonal muscles may pull one or more vertebrae out of line to one side, pressing nerves and producing additional pain to that from the trigger points.

Rotatores

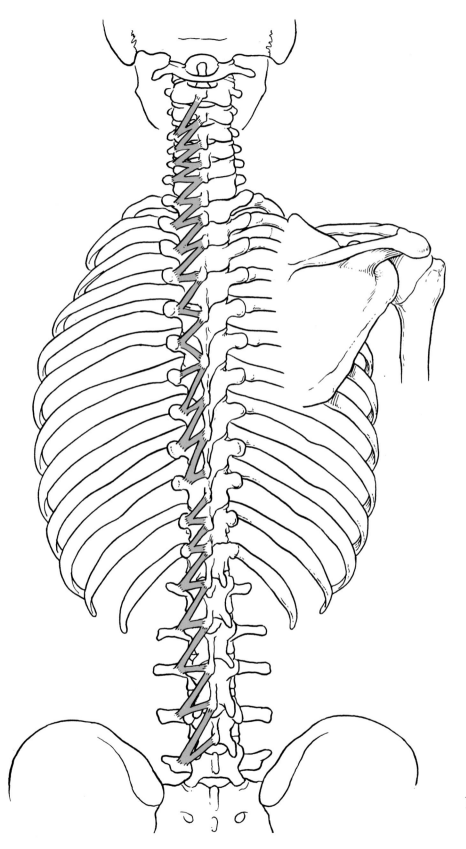

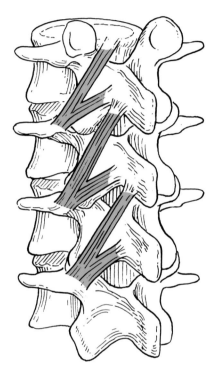

Origin: Transverse processes of each vertebrae

Insertion: Base of spinous process of next vertebrae above

Action: Extend and rotate the vertebral column

Innervation: Dorsal rami of spinal nerves

Interspinales

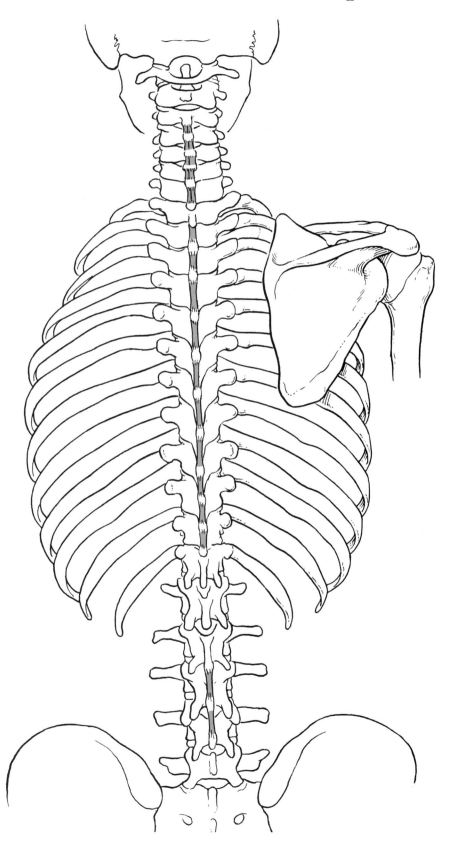

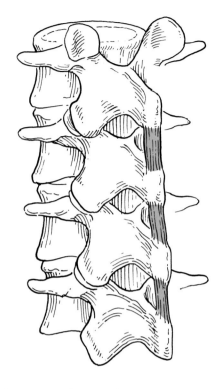

Origin: Cervical region—spinous processes of third to seventh cervical vertebrae (C3–C7)

Thoracic region—spinous processes of second to twelfth thoracic vertebrae (T2–T12)

Lumbar region—spinous processes of second to fifth lumbar vertebrae (L2–L5)

Insertion: Spinous process of next superior vertebra to the vertebra of origin

Action: Extend the vertebral column

Innervation: Posterior primary rami of spinal nerves

Intertransversarii

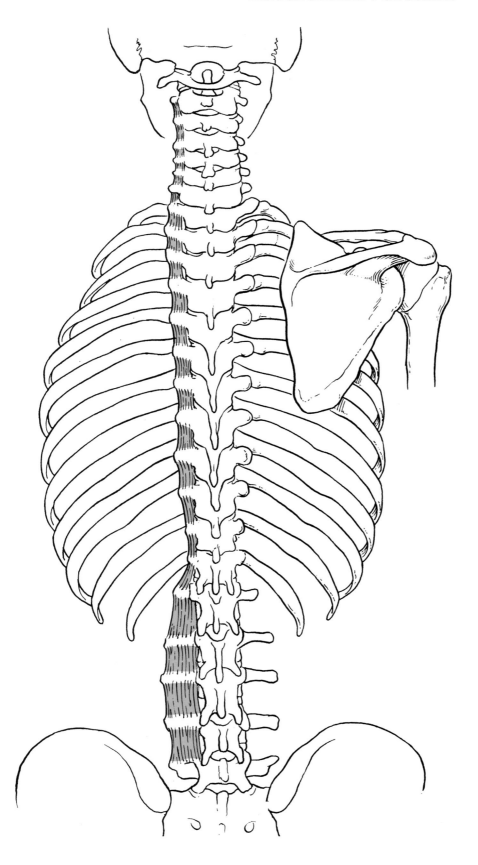

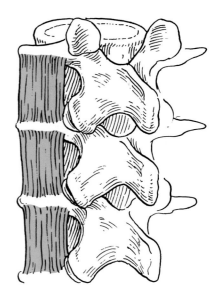

Origin: Transverse processes of all vertebrae from lumbar to axis

Insertion: Transverse process of next superior vertebrae

Action: Lateral flexion of vertebral column

Innervation: Ventral and dorsal rami of spinal nerves

Intercostales Externi

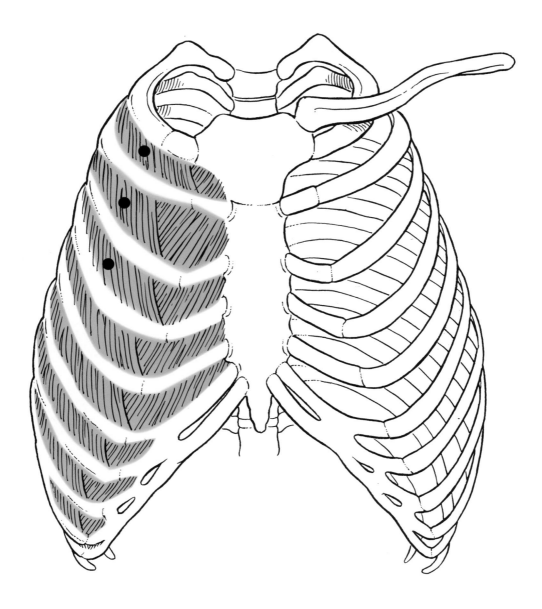

Origin: Lower margin of upper eleven ribs

Insertion: Superior border of rib below

Action: With first ribs fixed by scalenes, they pull the ribs toward one another to elevate the rib cage.

Innervation: Intercostal nerves

The **external intercostal** muscles act synergistically with the diaphragm to aid in inspiration. The fibers are oriented obliquely down and forward towards the costal cartilages. In the lower intercostal spaces the fibers are continuous with the **external oblique** muscle of the abdominal wall. The muscles can be **palpated** between the ribs. The **trigger points** are located anteriorly between the ribs.

Intercostales Interni

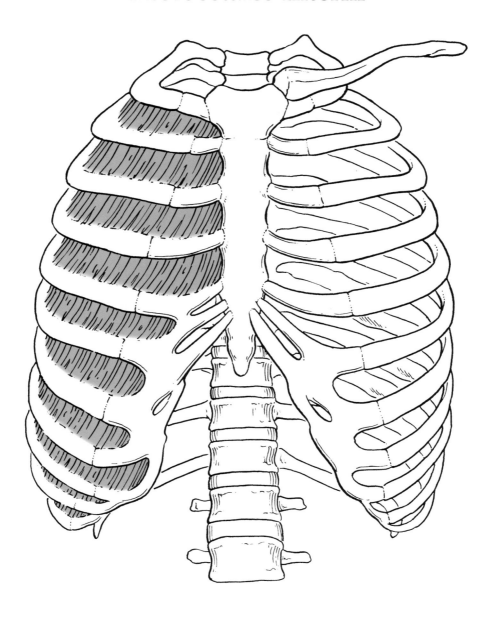

Origin: Ridge of inner surface of rib and corresponding costal cartilage

Insertion: Superior border of rib below

Action: Draw ribs together and depress the rib cage

Innervation: Intercostal nerves

The muscle fibers here are angled obliquely away from the costal cartilages. The contraction of these muscles decreases the size of the thoracic cavity and aids in forced expiration. The internal intercostals are antagonistic to the external intercostals.

Subcostales

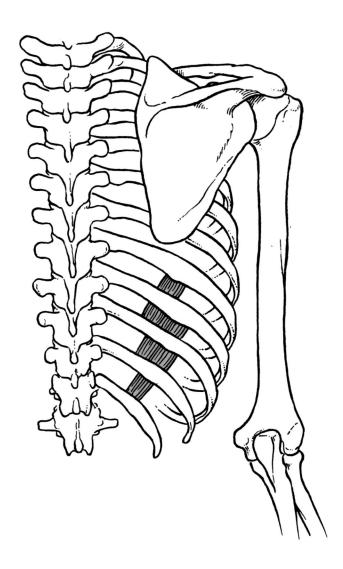

Origin: Inner surface of each rib near its angle

Insertion: Medially on the inner surface of the second or third rib below

Action: Draws ventral part of ribs downward

Innervation: Intercostal nerves

These muscles are deep to the internal intercostals. They are synergistic with the internal intercostals in decreasing the size of the thoracic cavity and aiding in forced expiration.

Transversus Thoracis

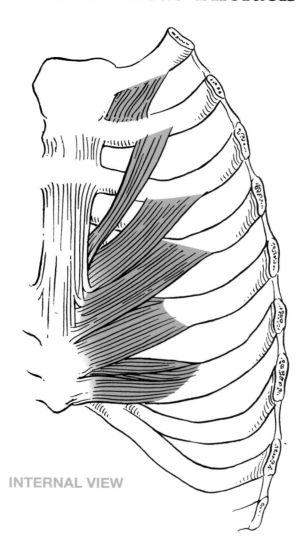

INTERNAL VIEW

Origin: Inner surface of the body of the sternum, xiphoid process, and sternal ends of the costal cartilages of the last three or four true ribs

Insertion: Inner surfaces of the costal cartilages of the second through sixth ribs

Action: Draws ventral part of rib downward

Innervation: Intercostal nerves

This muscle, found on the inside of the rib cage, decreases the size of the **thoracic cavity** and thus aids in forced expiration.

Levatores Costarum

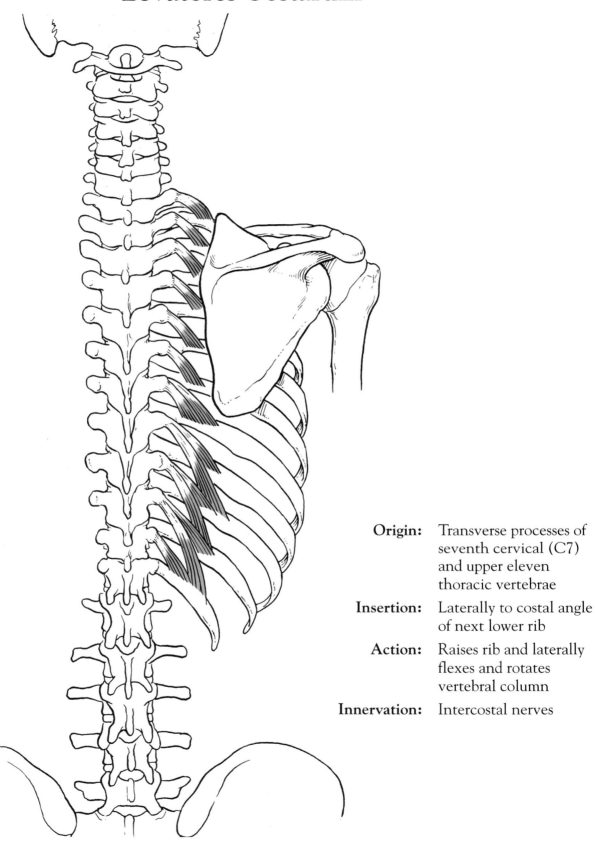

Origin: Transverse processes of seventh cervical (C7) and upper eleven thoracic vertebrae

Insertion: Laterally to costal angle of next lower rib

Action: Raises rib and laterally flexes and rotates vertebral column

Innervation: Intercostal nerves

Serratus Posterior Superior

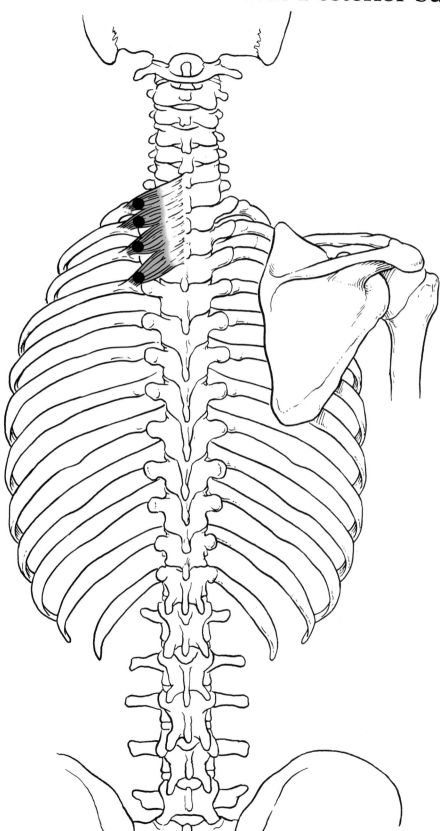

Origin: Lower portion of ligamentum nuchae and the spinous processes of the sixth and seventh cervical through the third thoracic vertebrae (C7–T3)

Insertion: Upper borders and external surfaces of ribs two through five lateral to their angles

Action: Assists in raising ribs during inspiration

Innervation: Second through fifth intercostal nerves (I2–I5)

This muscle lies under the **rhomboideus** next to the ribs. The **trigger points** are under the scapula near the insertion of the muscle on the ribs. Its **referred pain pattern** is under the upper portion of the scapula.

Serratus Posterior Inferior

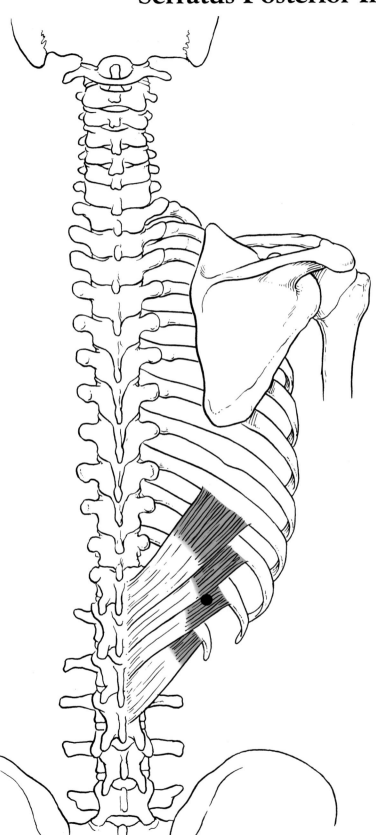

Origin: Spinous process of the last two thoracic and upper three lumbar vertebrae

Insertion: Inferior borders and outer surfaces of lower four ribs just lateral to the angles

Action: Depresses last four ribs (this is somewhat controversial in light of recent studies since it shows no electro-myographic activity during respiration)

Innervation: The ninth through twelfth spinal nerves (S9–S12)

The **trigger point** is the belly of the muscle near the eleventh rib. Its **referred pain pattern** is a nagging ache in the area of the muscle.

Diaphragm

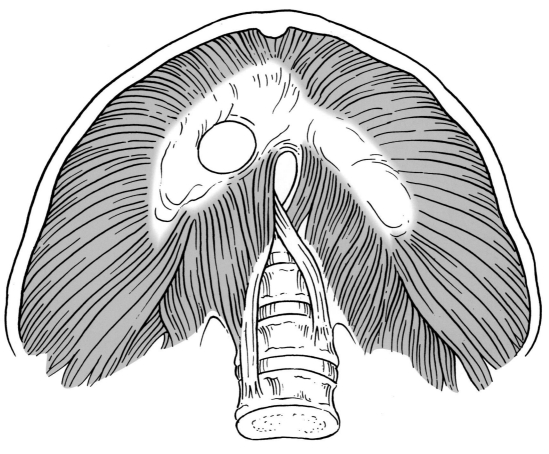

INFERIOR VIEW

Origin: First three lumbar vertebrae, lower six costal cartilages, and inner surface of xiphoid process of sternum

Insertion: Muscle fibers converge upward and inward to form the central tendon

Action: Flattens on contraction increasing the vertical dimensions of thorax

Innervation: Phrenic nerve

The **diaphragm** is the most important muscle of inspiration. When the muscle is relaxed it is dome shaped. It flattens as it contracts increasing the volume of the thoracic cavity. The alternate contraction and relaxation causes pressure changes in the abdominopelvic cavity that assist in the return of venous blood and lymph to the heart. The **Heimlich Maneuver** causes pressure on the diaphragm to increase intrathoracic pressure that forces food out of the laryngeal opening of a choking victim.

Rectus Abdominis

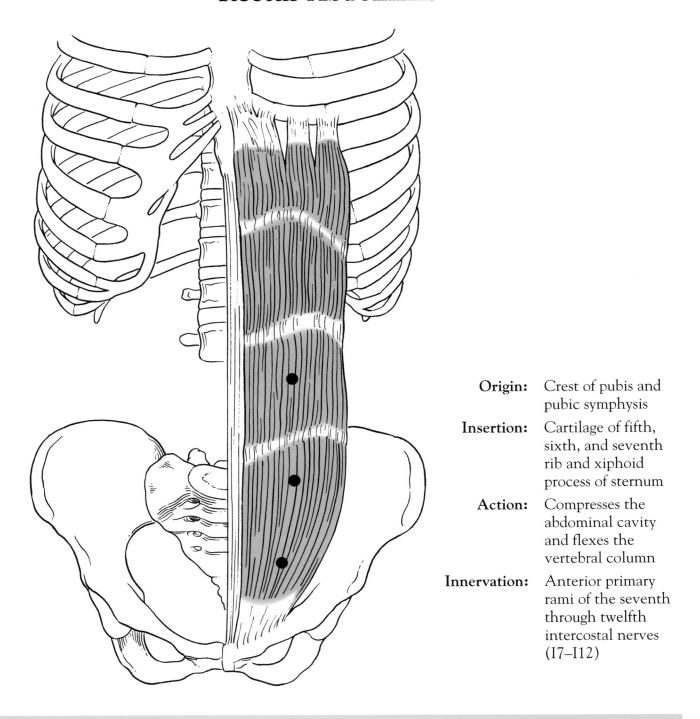

Origin: Crest of pubis and pubic symphysis

Insertion: Cartilage of fifth, sixth, and seventh rib and xiphoid process of sternum

Action: Compresses the abdominal cavity and flexes the vertebral column

Innervation: Anterior primary rami of the seventh through twelfth intercostal nerves (I7–I12)

These are the "abs". Tendinous bands divide each **rectus** into four bellies. Each muscle is enclosed in a sheath formed from the aponeurosis of the lateral abdominal muscles. It can be **palpated** on the anterior medial surface of the abdomen during active flexion of the trunk. This muscle contracts strongly during sit ups or when a person is lying in a supine position and raises the legs several inches from the floor. The **trigger points** are located below the umbilicus near the linea alba.

Obliquus Externus Abdominis

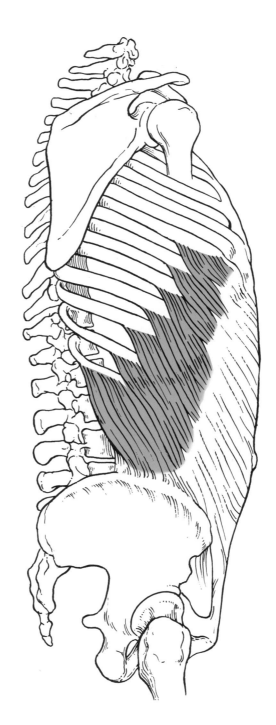

Origin: External surface of the lower eight ribs

Insertion: Anterior part of iliac crest and by abdominal aponeurosis to linea alba

Action: Compresses the abdominal cavity. Laterally flexes and rotates vertebral column. Both sides together flex vertebral column anteriorly.

Innervation: Ventral rami of the lower six thoracic nerves (T7–T12)

This is the most superficial of the three side abdominal muscles. Its fibers angle obliquely downward and medially. It can be **palpated** on the lateral sides of the abdomen during active trunk rotation to the opposite side.

Obliquus Internus Abdominis

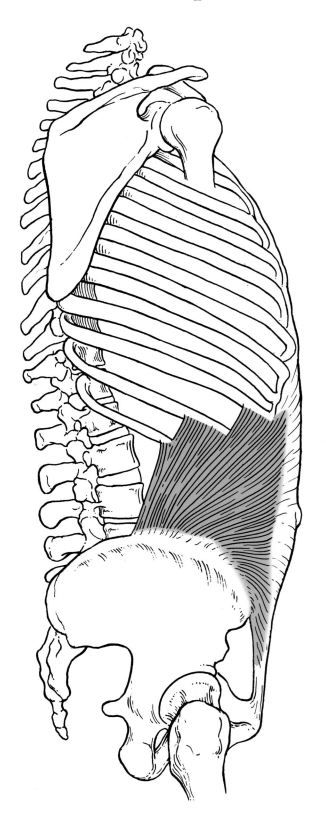

Origin: Lateral half of inguinal ligament, anterior two thirds of the iliac crest, and thoracolumbar fascia

Insertion: Upper fibers into cartilages of last three ribs, the remainder into the aponeurosis extending from the tenth costal cartilage to the pubic bone

Action: Compresses abdominal contents, laterally bends and rotates vertebral column. It also aids the rectus abdominus in flexing vertebral column.

Innervation: Ventral rami of the lower six thoracic and first lumbar spinal nerves (T7–T12, L1)

The **obliquus internus abdominis** is important in forced expiration, coughing and sneezing. Contraction squeezes the abdominal contents. It is the middle of the three layers of abdominal wall muscles.

Transversus Abdominis

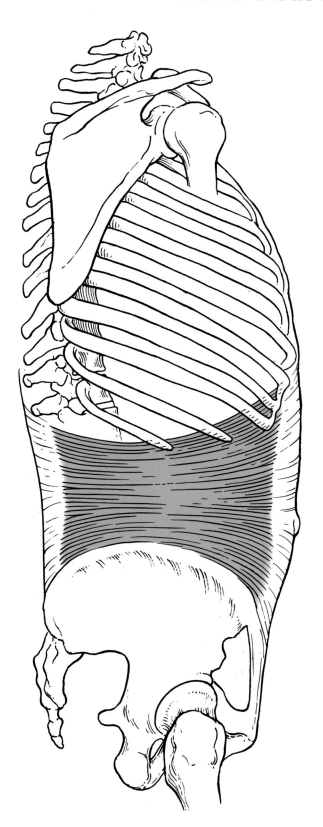

Origin: Lateral part of inguinal ligament, iliac crest, thoracolumbar fascia, and cartilage of lower six ribs

Insertion: Abdominal aponeurosis to linea alba

Action: Constricts the abdomen and supports the abdominal viscera

Innervation: Ventral rami of the lower six thoracic and first lumbar spinal nerves (T7–T12, L1)

The **transversus abdominis** is the innermost of the three abdominal muscle layers. Its fibers run horizontally while the other two abdominal muscle layers' fibers run obliquely.

Cremaster

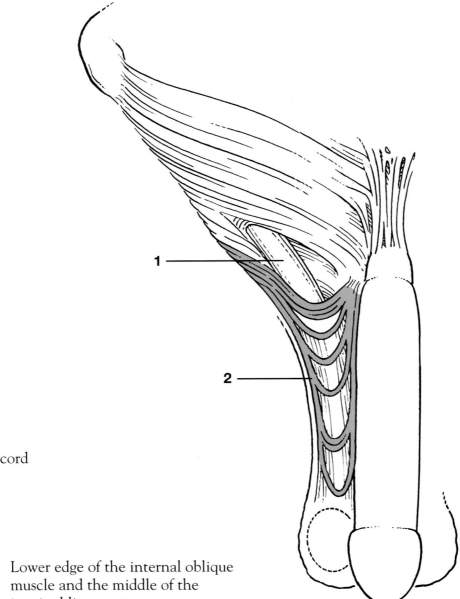

1. Spermatic cord
2. Cremaster

Origin: Lower edge of the internal oblique muscle and the middle of the inguinal ligament

Insertion: Pubic tubercle and crest of pubis

Action: Pulls the testes up toward the superficial inguinal ring

Innervation: Genital branch of the genito-femoral nerve from the first and second lumbar nerves (L1, L2)

Contraction of the **cremaster muscle** reflexively elevates the testis to a higher position in the scrotum for warmth and to protect against injury. Under very warm conditions, the muscles relax, enabling the testis to sit lower with greater heat loss from the surrounding scrotal skin. These responses help maintain an optimal temperature in the testis for the production of male sex cells. Although a striated muscle, the cremaster is not usually under voluntary control.

Quadratus Lumborum

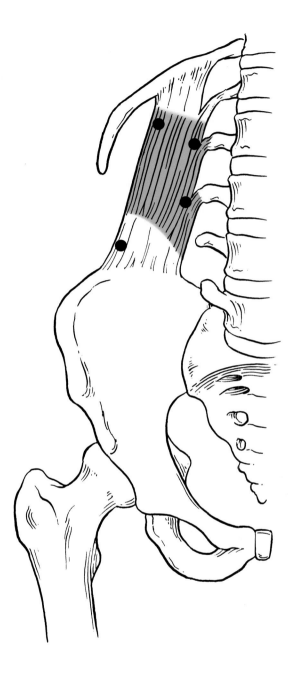

Origin: Iliolumbar ligament and the posterior portion of the iliac crest

Insertion: Inferior border of last rib and the transverse processes of the first four lumbar vertebrae (L1–L4)

Action: Flexes lumbar region of vertebral column laterally to the same side. Both muscles together stabilizes and extends the lumbar vertebrae and assists forced expiration.

Innervation: Ventral rami of the twelfth thoracic (T12) and upper three lumbar spinal nerves (L1–L3)

This muscle can be **palpated** in the supine position by palpating deep in the lumbar region above the iliac crest during active elevation of the hip. Its **trigger points** are found laterally near the rib or iliac attachment. Its **referred pain pattern** is found in the gluteal and groin area and also in the sacroiliac joint and the greater trochanter. If the trigger points are active, a cough or sneeze can cause severe pain in the lower back.

Superficial Transverse Perineus

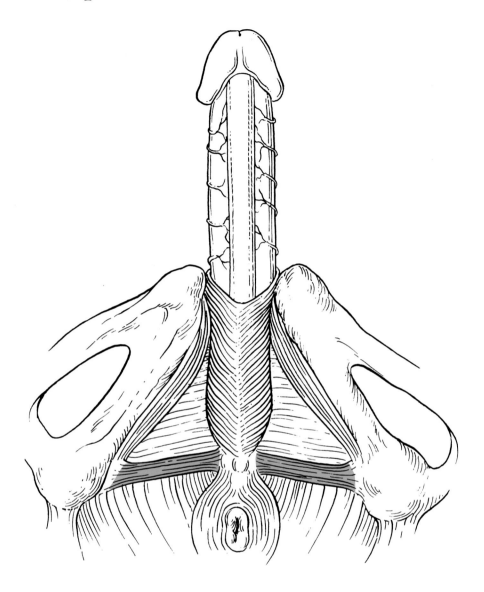

Action: Medial and anterior part of the ischial tuberosity

Insertion: Central tendinous point of perineum

Action: Stabilizes and strengthens perineum

Innervation: Perineal branches of pudendal nerve

These are paired muscle bands posterior to the **urethral**, and in females **vaginal**, openings. They are sometimes absent.

Coccygeus

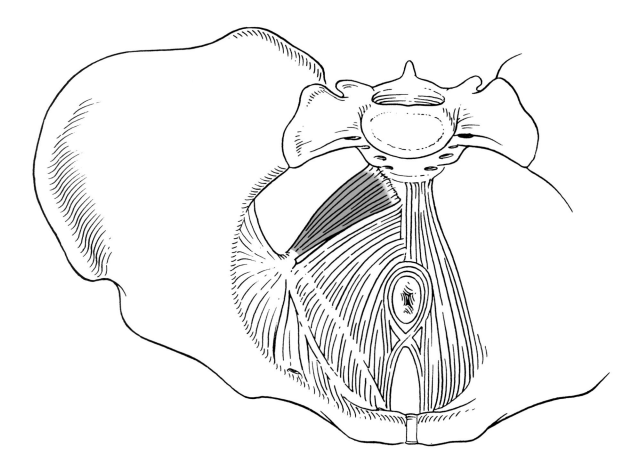

Origin: Pelvic surface of the ischial spine and the sacrospinous ligament

Insertion: Margin of the coccyx and the lower sacrum

Action: Supports pelvic viscera, supports coccyx and pulls it forward after it has been reflected by defecation or childbirth, and assists in closing the posterior part of the pelvic outlet

Innervation: Fourth and fifth sacral nerves (S4, S5)

This is a small triangular muscle lying posterior to the **levator ani**. It forms the posterior part of the pelvic diaphragm.

Sphincter Urethrae

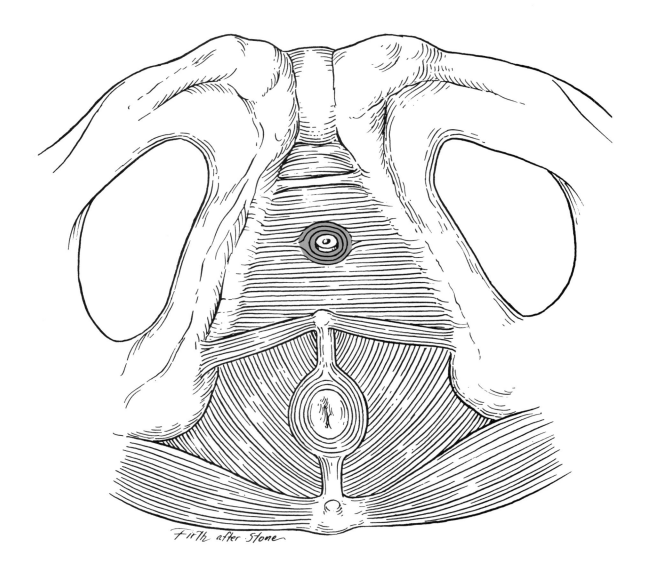

Firth after Stone

Origin: Ischiopubic rami

Insertion: Midline raphe

Action: Constricts urethra and helps support pelvic organs

Innervation: Perineal branch of pudendal nerve

In the female, this muscle encircles the **urethra** and the **vagina**.

Levator Ani

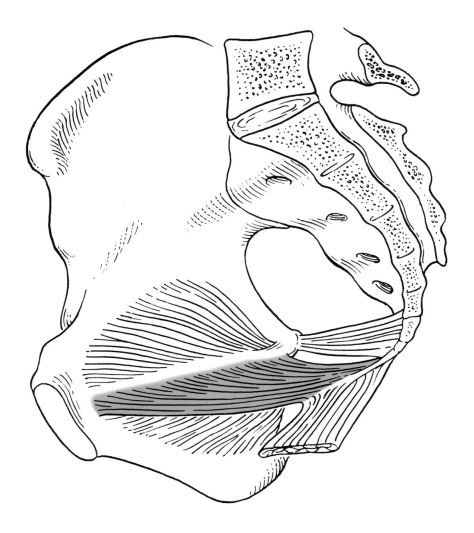

Origin: Pelvic surfaces of the pubis, inner surface of the ischial spine and the obturator fascia

Insertion: Inner surface of the coccyx, levator ani of opposite side, and sides of rectum

Action: Forms the floor of the pelvic cavity, constricts the lower end of the rectum and vagina, and supports and slightly raises the pelvic floor

Innervation: Fourth sacral nerve and inferior rectal nerve (S4)

This is a broad, flat muscle whose fibers extend inferomedially forming a muscular sling around the male **prostate** or female **vagina**, **urethra** and **anorectal junction** before meeting in the median plane. It can be **palpated** internally in the female.

Ischiocavernosus

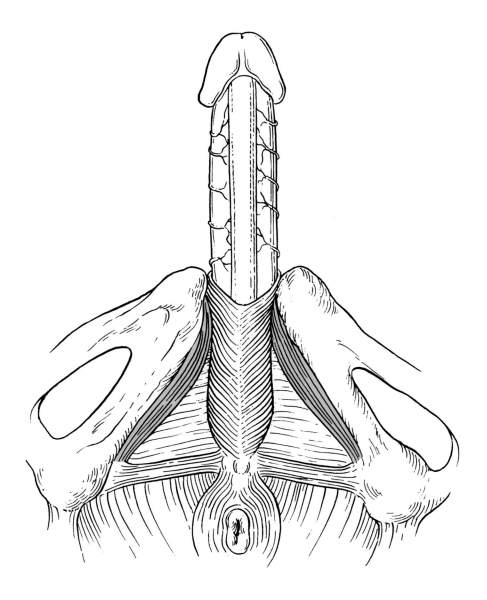

Origin: Inner surface of the ischial tuberosity and the ramus of the ischium

Insertion: Aponeurosis on the sides and undersurface of the crus penis or clitoris

Action: Compresses the crus penis which obstructs venous return and maintains erection of penis or clitoris

Innervation: Perineal branch of the pudendal nerve

External Sphincter Ani

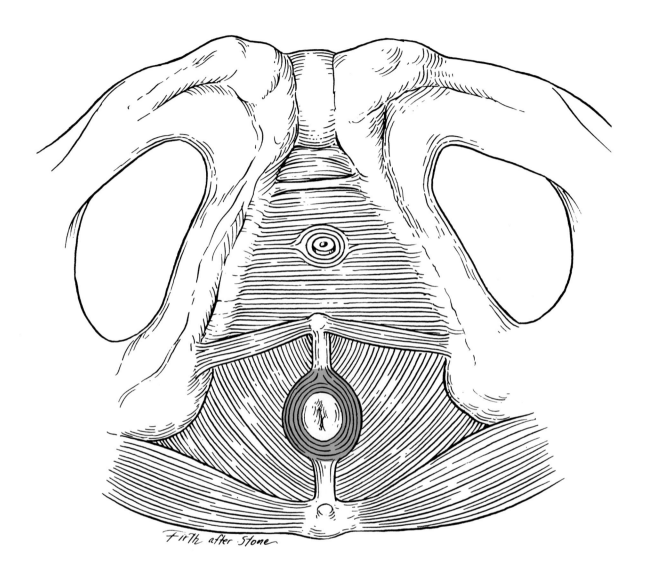

Firth after Stone

Origin:	Central tendon of perineum
Insertion:	Midline raphe and the coccygeus muscle
Action:	Voluntary muscle circling the anus preventing defecation
Innervation:	Pudendal nerve

Bulbospongiosus

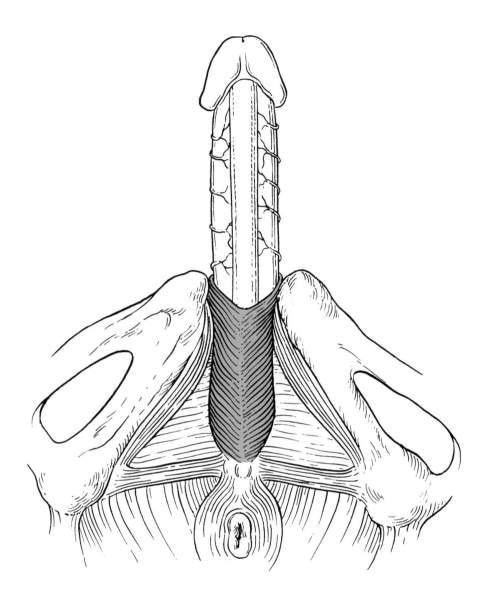

Origin:	Central tendon of the perineum and midline raphe of male penis
Insertion:	Anteriorly into corpus cavernosa of penis or clitoris.
Action:	Empties male urethra and assists in erection of penis in male and clitoris in female
Innervation:	Perineal branch of pudendal nerve

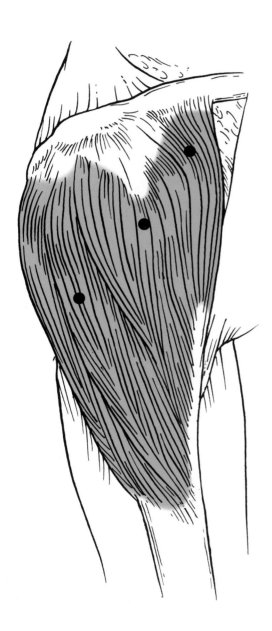

7

Muscles of the Shoulder and Upper Arm

Pectoralis Major

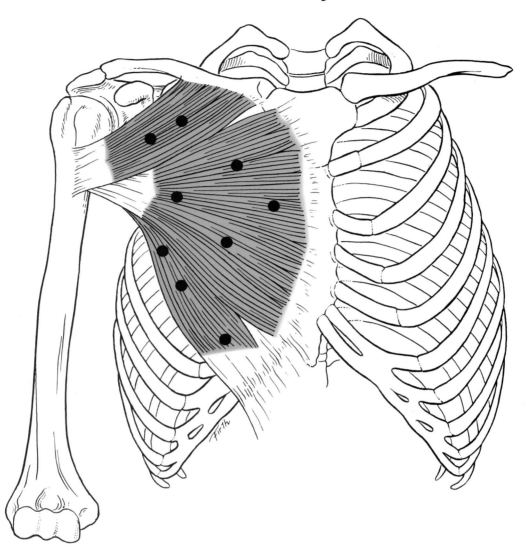

Origin: Ventral surface of the sternum down to the seventh rib, sternal half of clavicle, cartilage of true ribs and aponeurosis of the external oblique muscle

Insertion: Lateral lip of the intertubercular groove of the humerus

Action: Adducts and medially rotates arm, clavicular head flexes humerus, sternal head extends humerus, and with insertion fixed it assists in elevation of the thorax

Innervation: Medial and lateral pectoral nerves (C5–C8, T1))

The **trigger points** for this muscle are in the belly for each portion, and the **referred pain pattern** is the chest and breast down to the ulnar aspect of the arm to the fourth and fifth fingers. It can be palpated along the anterior border of the axilla during active adduction of the humerus. In "pull ups", it pulls the thorax up to the fixed arm position.

Pectoralis Minor

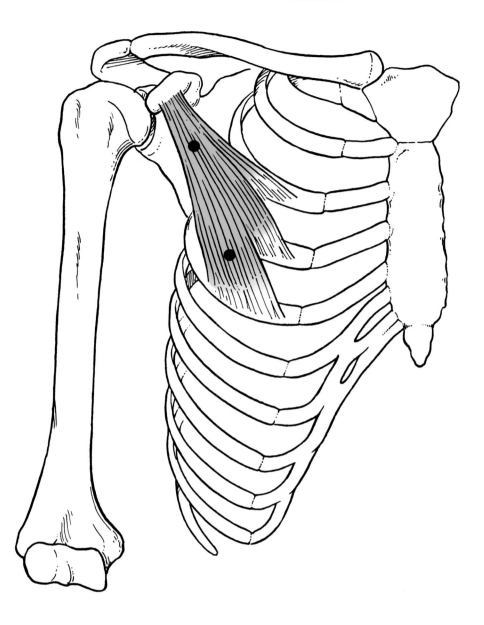

Origin: Anterior surfaces of the third through fifth rib

Insertion: Coracoid process of the scapula

Action: With ribs fixed, it draws the scapula forward and downward, and with scapula fixed, it draws the rib cage superiorly

Innervation: Medial pectoral nerve (C8, T1))

The **trigger points** for this muscle are near the insertion at the ribs and at the coracoid process. Its **referred pain pattern** is the front of the chest and down the ulnar side of the arm and mimics the symptoms of angina. It is difficult to **palpate**.

Subclavius

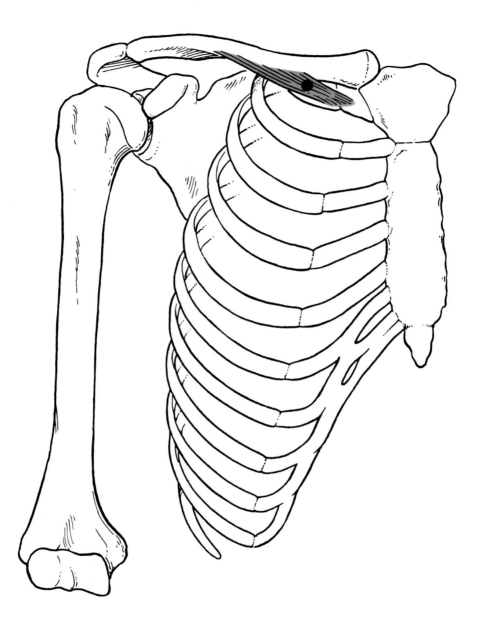

Origin: Junction of the first rib and its costal cartilage

Insertion: Groove on the inferior surface of clavicle

Action: Depresses clavicle and draws shoulder forward and downward

Innervation: C5, C6

The **trigger point** is in the belly of the muscle.

Coracobrachialis

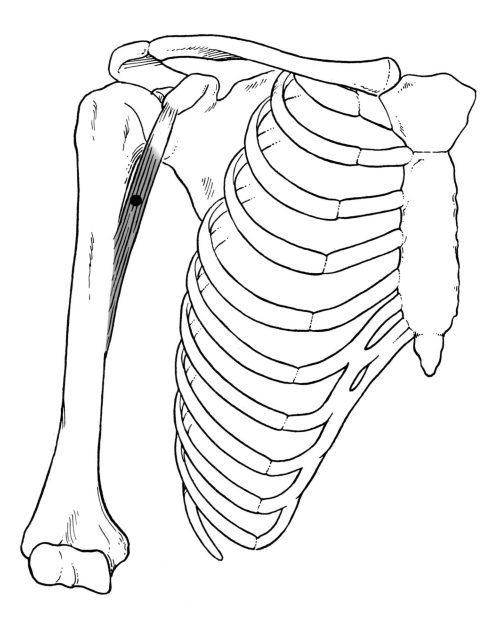

Origin: Tip of coracoid process of scapula

Insertion: Anteromedial surface of the humerus shaft

Action: Flexion and adduction of the humerus

Innervation: Musculocutaneous nerve (C5–C7)

The **trigger point** for this muscle is near the coracoid attachment, and its **referred pain pattern** is down the triceps and dorsal forearm into the hand. Although it is difficult to palpate, it can be **palpated** medial to the short head of the biceps brachii when the humerus is flexed against resistance.

Brachialis

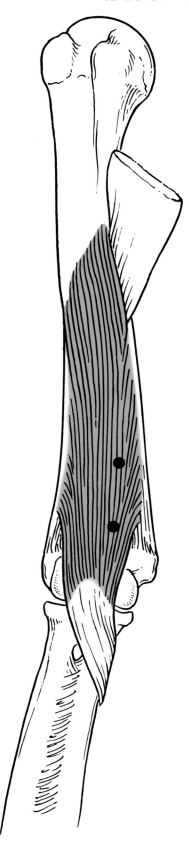

Origin: Distal half of the anterior surface of the humerus

Insertion: Coronoid process and tuberosity of ulna

Action: Flexes forearm

Innervation: Musculocutaneous and radial nerves (C5, C6)

This muscle can be palpated medial to the **biceps brachii** on lower anterior humerus during active flexion of the elbow. Its tendon can be felt deep in antecubital fossa just medial to the tendon of the biceps. The **trigger points** are found in the belly of the muscle. Its **referred pain pattern** is primarily in the lower arm to the thumb. This muscle is a strong elbow flexor.

Biceps Brachii

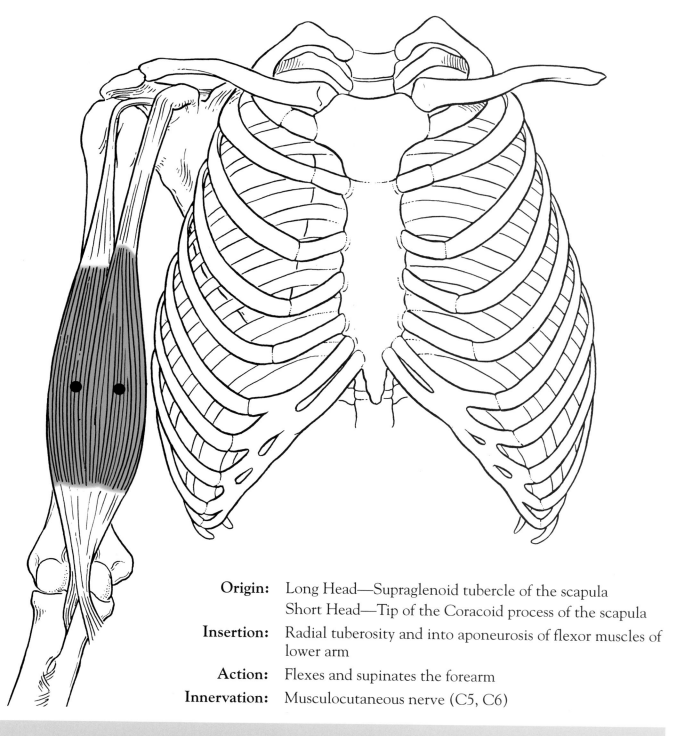

Origin:	Long Head—Supraglenoid tubercle of the scapula
	Short Head—Tip of the Coracoid process of the scapula
Insertion:	Radial tuberosity and into aponeurosis of flexor muscles of lower arm
Action:	Flexes and supinates the forearm
Innervation:	Musculocutaneous nerve (C5, C6)

This muscle can readily be **palpated** on the anterior surface of the humerus. Its tendon can also be palpated during active elbow flexion in the antecubital fossa. The **trigger points** are found in the belly of each of the two heads of the muscle. Its **referred pain pattern** is found in the front of the shoulder and into the antecubital space (the "crease" of the elbow).

Latissimus Dorsi

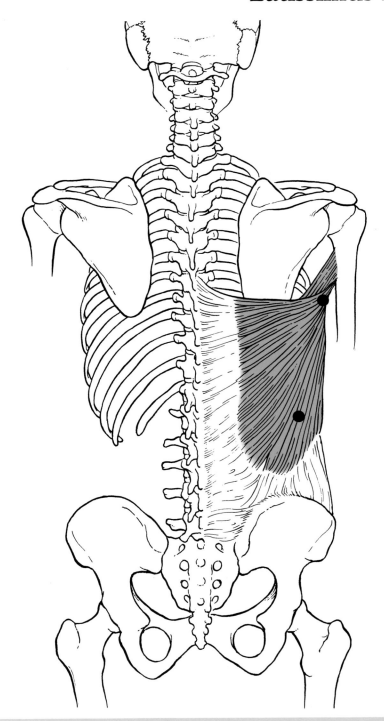

Origin: Indirect attachment through lumbodorsal fascia into spinous process of lower six thoracic and lumbar vertebrae, lower three to four ribs and iliac crest

Insertion: In floor of intertubercular groove of humerus

Action: Extends, adducts and medially rotates the arm; draws the shoulder downward and backward

Innervation: Thoracodorsal nerve (C6–C8)

This muscle is important in bringing the arm down in a power stroke as in hammering, swimming (crawl stroke), and rowing. Its **trigger points** are in the belly of the muscle near the rib attachments. Its **referred pain pattern** is below the scapula and into the ulnar side of the arm and the abdominal oblique area. It can be **palpated** along the lateral side of the rib cage during active extension of the humerus. It also acts as an accessory muscle of respiration.

Latissimus Dorsi

The latissimus dorsi as a possible cardiovascular assist muscle

Although there are effective mechanical heart assist devices today, in the mid 1980s two surgical techniques were developed for using the **latissimus dorsi** muscle to compensate for weak cardiac muscle contraction. The **latissimus** (Figure 1) was cut (Figure 2) and the anterior portion used either to wrap around the heart (Figure 3) or to form a pouch (lined by GORE-TEX) into which the cut ends of the dorsal aorta (with artificial GORE-TEX connections) were diverted (Figure 4). A mechanical pacemaker stimulated contraction of the muscle wrapped around the ventricle or forming the pouch in coordination with the heart beat to increase the force with which the blood is propelled through the circulatory system. Advantages of both techniques included no risk of rejection or infection (the only 'artificial' component was the GORE-TEX and pacemaker, both compatible with body tissues), freedom from an external power source and no time loss seeking heart donors.

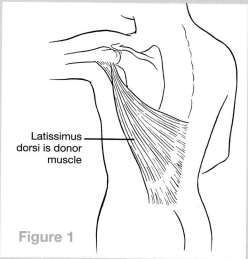

Latissimus dorsi is donor muscle

Figure 1

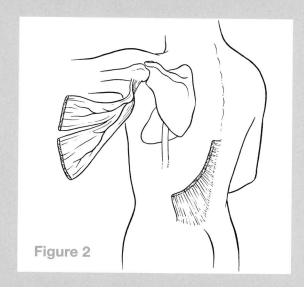

Figure 2

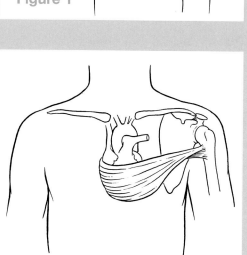

Figure 3

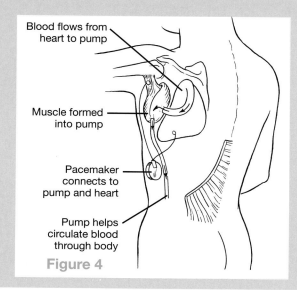

Blood flows from heart to pump

Muscle formed into pump

Pacemaker connects to pump and heart

Pump helps circulate blood through body

Figure 4

Trapezius

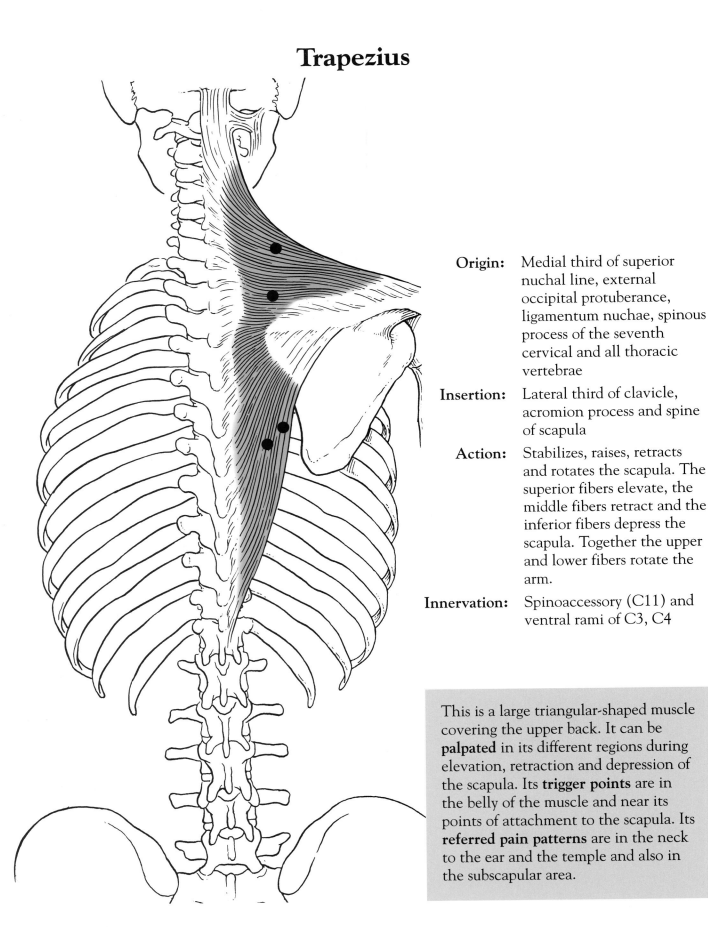

Origin: Medial third of superior nuchal line, external occipital protuberance, ligamentum nuchae, spinous process of the seventh cervical and all thoracic vertebrae

Insertion: Lateral third of clavicle, acromion process and spine of scapula

Action: Stabilizes, raises, retracts and rotates the scapula. The superior fibers elevate, the middle fibers retract and the inferior fibers depress the scapula. Together the upper and lower fibers rotate the arm.

Innervation: Spinoaccessory (C11) and ventral rami of C3, C4

This is a large triangular-shaped muscle covering the upper back. It can be **palpated** in its different regions during elevation, retraction and depression of the scapula. Its **trigger points** are in the belly of the muscle and near its points of attachment to the scapula. Its **referred pain patterns** are in the neck to the ear and the temple and also in the subscapular area.

An Illustrated Atlas of the Skeletal Muscles

Levator Scapulae

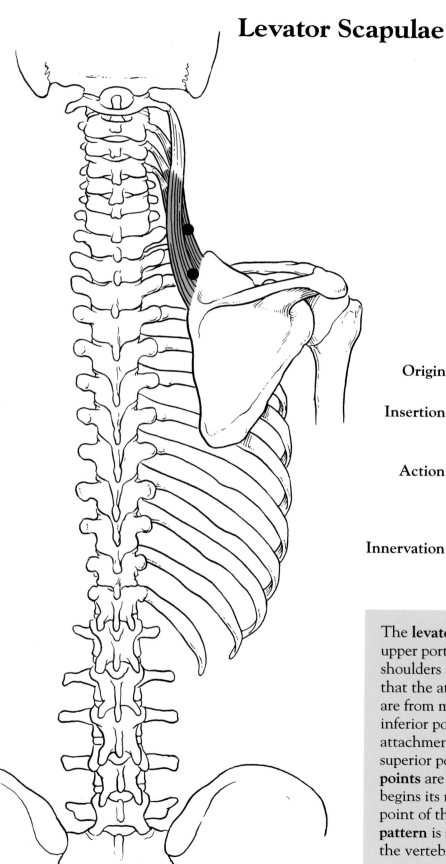

Origin: Transverse processes of the first four cervical vertebrae

Insertion: Vertebral border of the scapula between the superior angle and the spine

Action: Raises scapula and draws it medially. With the scapula fixed, bends the neck laterally and rotates it to the same side

Innervation: Third and fourth cervical spinal nerves and dorsal scapular nerve (C5)

The **levator** is an elevator. It works with the upper portion of the **trapezius** when the shoulders are shrugged. It has a twist in it so that the attachments at the atlas and axis are from muscle fibers that attach to the inferior portion of the border and the attachment at C4 are from fibers at the superior portion of the border. Its **trigger points** are at the belly of the muscle just as it begins its rotation and at the attachment point of the scapula. Its **referred pain pattern** is along the angle of the neck and the vertebral border of the scapula.

Rhomboideus Major

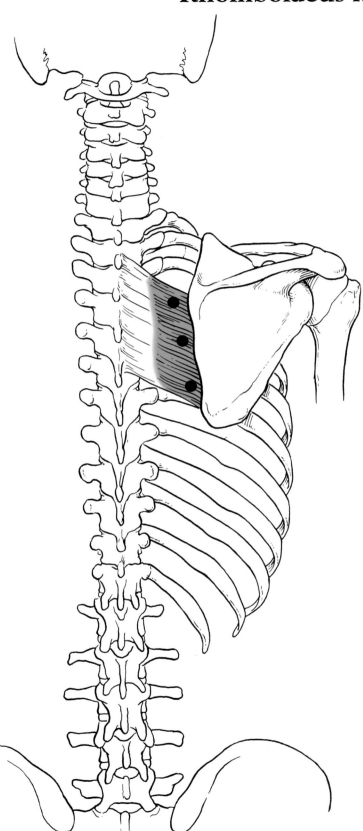

Origin: Spinous process of the second through fifth thoracic vertebrae

Insertion: Medial border of scapula between the spine and the inferior angle

Action: Adducts, retracts, elevates, and rotates the scapula so that the glenoid cavity faces downward and stabilizes the scapula

Innervation: Dorsal scapular nerve (C5)

This muscle can be **palpated** along its vertebral border during active scapular retraction. Its **trigger points** are found at the attachment point near the scapula border. Its **referred pain pattern** is in the scapula region.

Rhomboideus Minor

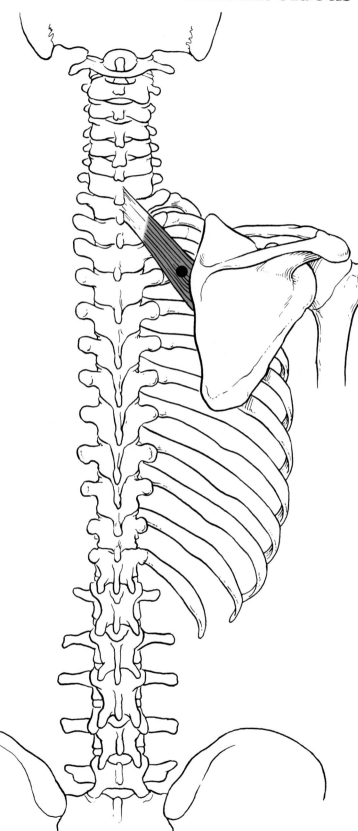

Origin: Spinous processes of the seventh cervical and first thoracic vertebrae

Insertion: Medial border of the scapula at the root of the spine

Action: Retracts and stabilizes the scapula, elevates the vertebral border, and rotates the scapula to depress the inferior angle

Innervation: Dorsal scapular nerve (C5)

The fibers of both **rhomboideus** muscles are arranged in an oblique, downward pattern. The **palpation** directions and **trigger points** are the same as for the **rhomboideus major.**

Serratus Anterior

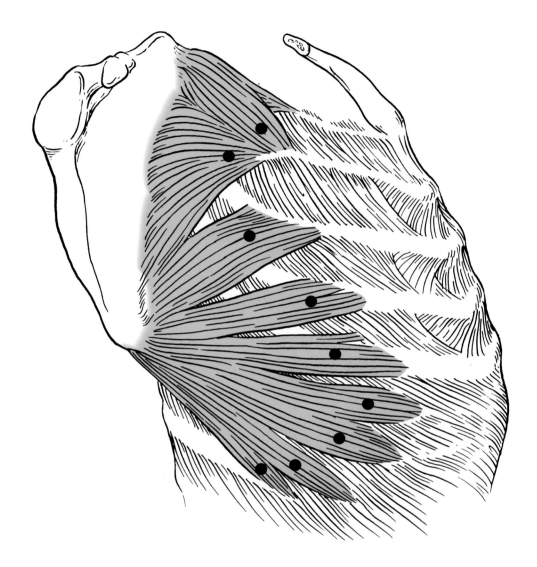

Origin: Outer surfaces and superior borders of the first eight or nine ribs

Insertion: Anterior surface of the medial border of the scapula

Action: Abducts and protracts the scapula, rotates it so that the glenoid cavity faces cranially. It also holds the scapula firmly against the thorax.

Innervation: Long thoracic nerve (C5–C7)

This muscle is important in horizontal arm movement such as pushing and punching. It is sometimes called the "boxer's muscle". It can be **palpated** along the anterior surface of the ribs below the axilla during active scapular protraction. Its **trigger points** are along the midaxillary line near the ribs. Its **referred pain pattern** is along the side and back of chest and down the ulnar aspect of the arm into the hand.

Subscapularis

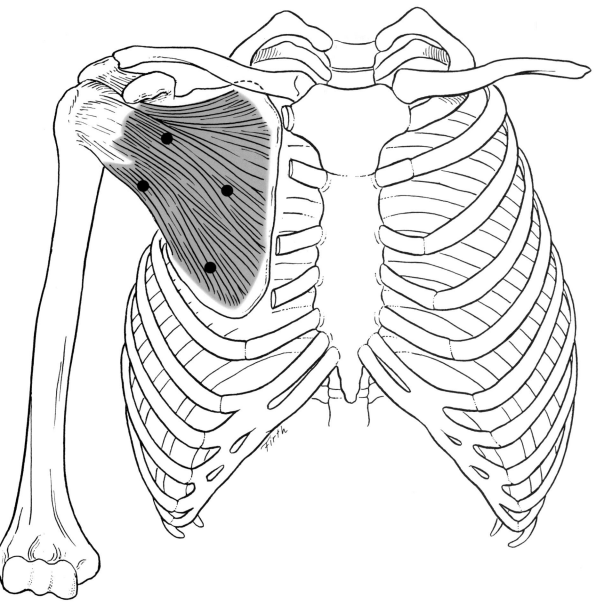

Origin: Subscapular fossa of the scapula.

Insertion: Lesser tubercle of the humerus and the ventral part of the capsule of the shoulder joint

Action: Medially rotates and stabilizes the head of the humerus in the glenoid cavity

Innervation: Upper and lower subscapular nerves (C5, C6)

This "rotator cuff" muscle is often implicated in "frozen shoulder syndrome". Because of its position it is difficult to **palpate**. Its **trigger points** are found near the attachment of the humerus and in its belly. Its **referred pain pattern** is in the posterior deltoid, triceps region down to the wrist.

Infraspinatus

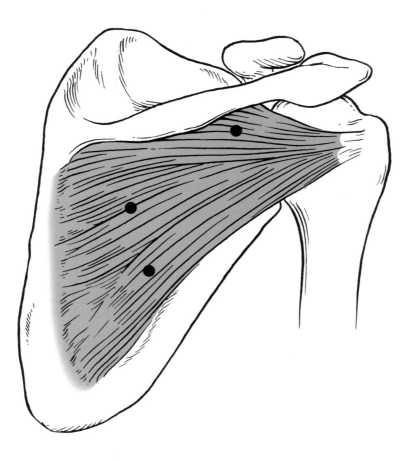

Origin: Infraspinous fossa of the scapula

Insertion: Middle facet of the greater tubercle of the humerus and the capsule of the shoulder joint

Action: Lateral rotation of the shoulder and acts to stabilize the humeral head in the glenoid cavity. It is one of the "rotator cuff" muscles.

Innervation: Suprascapular nerve (C5, C6)

This "rotator cuff" muscle has **trigger points** in the belly of the muscle below the spine of the scapula and near the medial border of the scapula. The **referred pain pattern** is deep into the shoulder and deltoid area down into the arm. This muscle can be **palpated** along the axillary border below the posterior deltoid during active lateral rotation of the humerus.

Supraspinatus

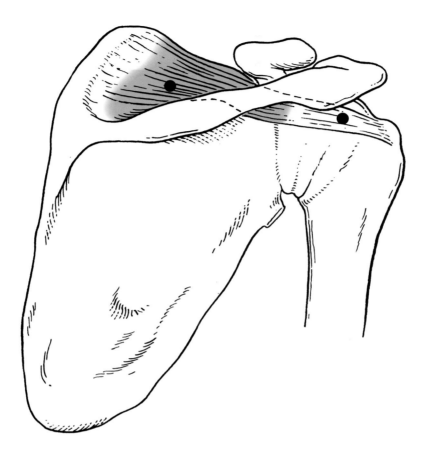

Origin: Supraspinous fossa of the scapula

Insertion: Superior facet of the greater tubercle of the humerus and the capsule of the shoulder joint

Action: Abducts the arm and acts to stabilize the humeral head in the glenoid cavity during movements of the shoulder joint. One of four "rotator cuff" muscles.

Innervation: Suprascapular nerve (C5, C6)

This is one of the four rotator cuff muscles, and it is the tendon of this muscle that frequently is torn. Pain is described as deep in the shoulder and becomes progressively worse during abduction. The **trigger points** are near the tendon and in the belly of the muscle. The **referred pain pattern** is deep in the shoulder and down the arm to the elbow. It can be **palpated** above the spine of the scapula during active abduction of the arm.

Teres Major

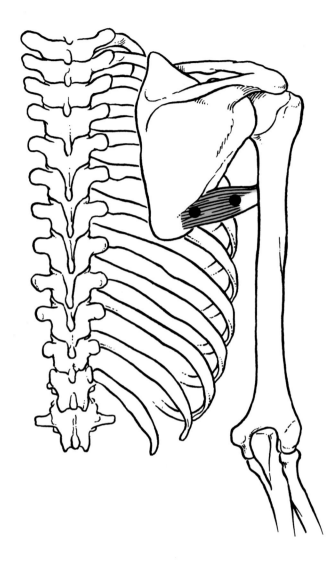

Origin: Lower third of the posterior surface of the lateral border of the scapula

Insertion: Medial lip of the bicipital groove of the humerus

Action: Medially rotates, adducts, and extends arm

Innervation: Lower subscapular nerve (C5, C6)

This muscle can be **palpated** from the inferior angle of the scapula diagonally upward during active extension of the humerus. Its **trigger points** are near both attachments. Its **referred pain pattern** is in the posterior deltoid region and down the dorsal surface of the arm.

Teres Minor

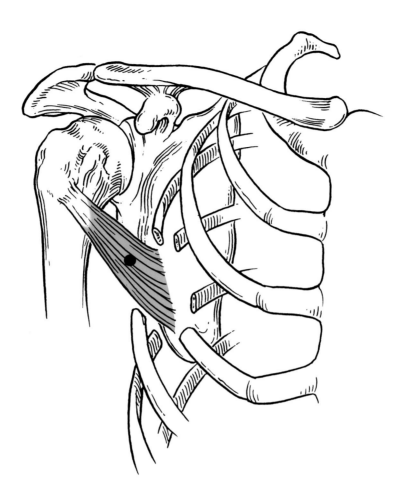

Origin: Upper two thirds of the dorsal surface of the axillary border of the scapula

Insertion: The capsule of the shoulder joint and the lower facet of the greater tubercle of the humerus

Action: Laterally rotates the arm and draws the humerus toward the glenoid cavity

Innervation: Axillary nerve (C5)

This is another one of the "rotator cuff" muscles. **Palpate** along the axillary border of the scapula immediately below the **infraspinatus** during active lateral rotation of the humerus. Its **trigger point** is in the belly of the muscle near its point of attachment. Its **referred pain pattern** is in the deltoid region.

Deltoideus

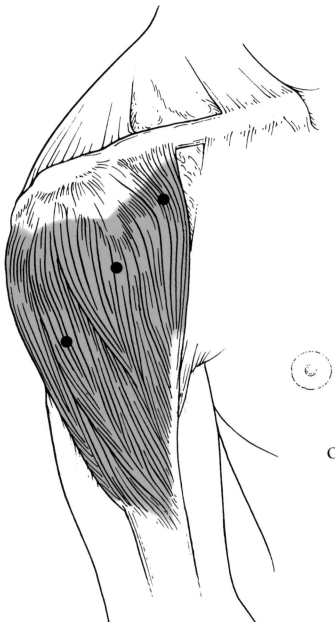

Origin: Anterior portion- superior surface of lateral third of clavicle

Middle portion—lateral border of acromion process of scapula

Posterior portion—lower border of the crest of the spine of the scapula

Insertion: Deltoid tuberosity of the humerus

Action: Anterior portion—flexion and medial rotation of the arm

Middle portion—abducts the arm

Posterior portion—extends and laterally rotates the arm

Innervation: Axillary nerve (C5, C6)

This shoulder muscle is one of the prime injection sites. It is active during the rhythmic arm swinging movements involved in walking. It can be **palpated** on the upper arm below the acromion process during movements of the humerus. Its **trigger points** are found in the belly of the muscle. Its **referred pain pattern** is in the deltoid region and down the lateral surface of the arm.

Triceps Brachii

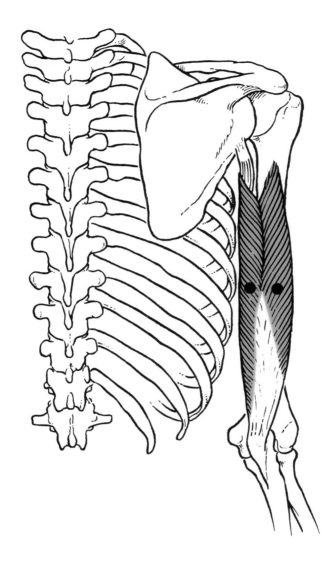

Origin: Long head—infraglenoid tubercle of the scapula

Medial head—distal two thirds of the medial and posterior surfaces of the humerus

Lateral head—upper half of the posterior surface of the humerus

Insertion: Posterior surface of the olecranon process of the ulna

Action: Extends the forearm and the tendon of the long head helps stabilize the shoulder joint

Innervation: Radial nerve (C7, C8)

The **triceps brachii** is the only muscle on the posterior upper arm. It can be **palpated** during active elbow extension. Its **trigger points** are in the belly of each head of the muscle. Its **referred pain pattern** is the entire length of the posterior surface of the arm.

Anconeus

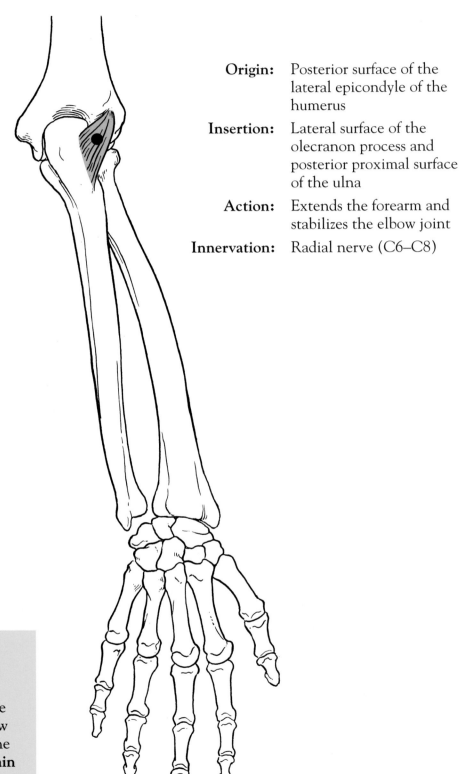

Origin: Posterior surface of the lateral epicondyle of the humerus

Insertion: Lateral surface of the olecranon process and posterior proximal surface of the ulna

Action: Extends the forearm and stabilizes the elbow joint

Innervation: Radial nerve (C6–C8)

This small muscle works with the **triceps brachii**. It can be **palpated** between the olecranon process of the ulna and the lateral epicondyle of the humerus during active elbow extension. Its **trigger point** is in the belly of the muscle. Its **referred pain pattern** is at the elbow.

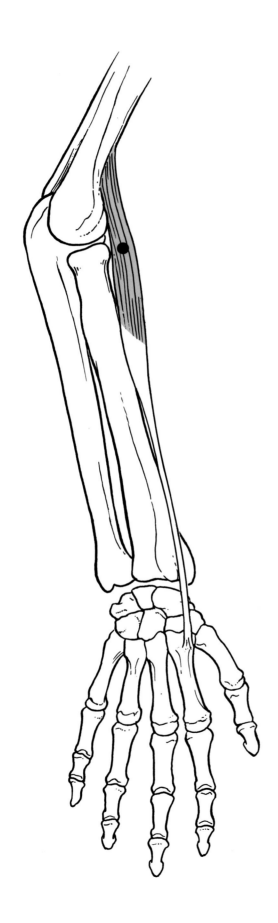

8

Muscles of the Forearm and Hand

Pronator Teres

Origin: Humeral head—just above the medial epicondyle of the humerus

Ulnar head—medial side of the coronoid process of the ulna

Insertion: Middle of lateral surface of radius

Action: Pronates the forearm and assists in flexing the elbow joint

Innervation: Median nerve (C6, C7)

This muscle can be **palpated** on the medial side of the forearm just medial to the insertion of the biceps during resisted pronation. It forms the medial border of the antecubital fossa. Its **trigger point** is the belly of the muscle near the elbow attachment. Its **referred pain pattern** is radial side of the forearm into the wrist and thumb.

Brachioradialis

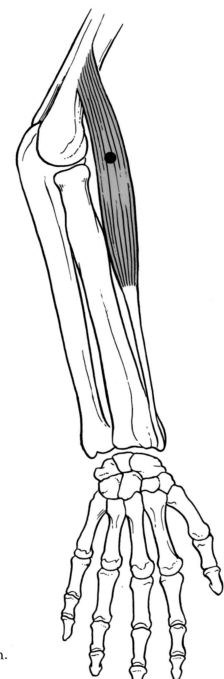

Origin: Proximal two thirds of the lateral supracondylar ridge of humerus.

Insertion: Lateral side of the base of the styloid process of radius

Action: Flexes the elbow. Assists in pronation and supination of the forearm to the midposition.

Innervation: Radial nerve (C5, C6)

The **brachioradialis** is used to stabilize the elbow during rapid flexion and extension while in a midposition, such as in hammering. The **trigger point** is in the belly of the muscle. The **referred pain pattern** is from the wrist and base of thumb in the web space between the thumb and index finger to the lateral epicondyle of the humerus. The muscle can be **palpated** on the upper forearm during resisted elbow flexion.

Flexor Carpi Radialis

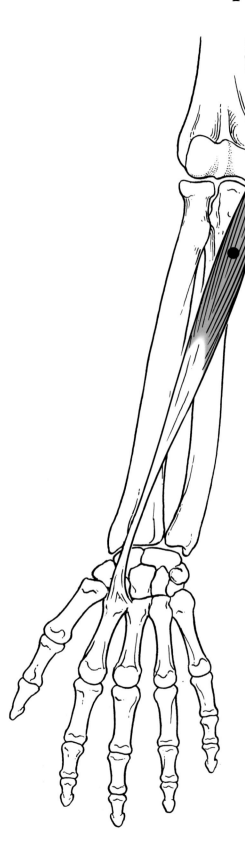

Origin: Medial epicondyle of the humerus

Insertion: Base of second and third metacarpal bones

Action: Flexes wrist and abducts hand

Innervation: Median nerve (C6, C7)

The **flexor carpi radialis** runs diagonally across the forearm. Its fleshy belly is replaced midway by a flat tendon that becomes cord-like at the wrist. The tendon is **palpated** on the anterior surface of the wrist in line with the second metacarpal. This tendon acts as a guide to the position of the radial artery in order to take a pulse. **Golfer's elbow** is a painful condition that follows repetitive use of the superficial muscles on the anterior forearm straining the **common flexor tendon.** The **flexor carpi radialis** is the most commonly affected muscle. Another name for this syndrome is **medial epicondylitis** reflecting the site of inflammation. The **trigger point** is in the belly of the muscle.

Flexor Carpi Ulnaris

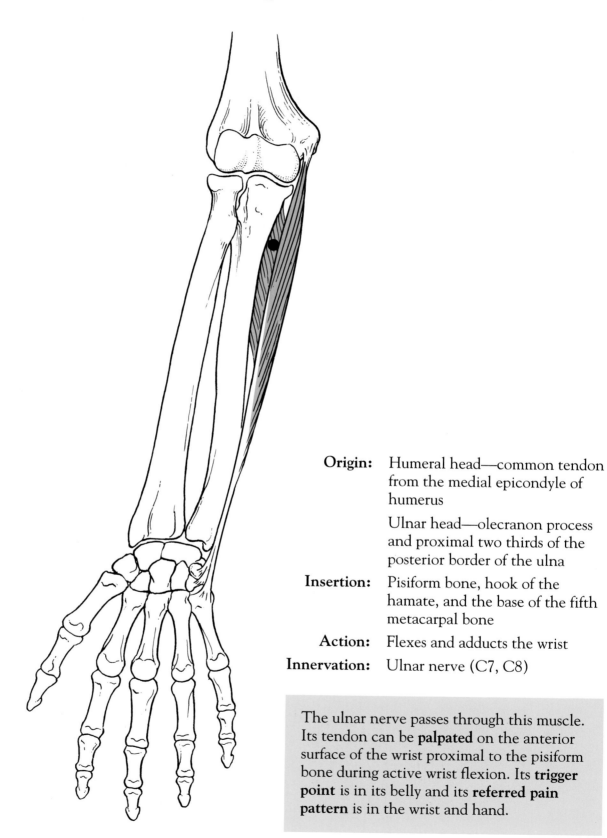

Origin: Humeral head—common tendon from the medial epicondyle of humerus

Ulnar head—olecranon process and proximal two thirds of the posterior border of the ulna

Insertion: Pisiform bone, hook of the hamate, and the base of the fifth metacarpal bone

Action: Flexes and adducts the wrist

Innervation: Ulnar nerve (C7, C8)

The ulnar nerve passes through this muscle. Its tendon can be **palpated** on the anterior surface of the wrist proximal to the pisiform bone during active wrist flexion. Its **trigger point** is in its belly and its **referred pain pattern** is in the wrist and hand.

Palmaris Longus

Origin: Medial epicondyle of the humerus through the common flexor tendon

Insertion: Front of flexor retinaculum and apex of the palmar aponeurosis

Action: Tenses the palmar fascia and flexes the wrist

Innervation: Median nerve (C6–C8)

The tendon of the **palmaris longus** is above the antebrachial fascia of the wrist and can be seen if one cups the hand and flexes the wrist. It can be **palpated** in midline of anterior surface of wrist when the wrist is flexed against resistance and the thumb is abducted. This muscle is absent in about one fourth of the population. Its **trigger point** is in the belly of the muscle. The **referred pain pattern** is into the wrist and fingers.

Flexor Digitorum Superficialis

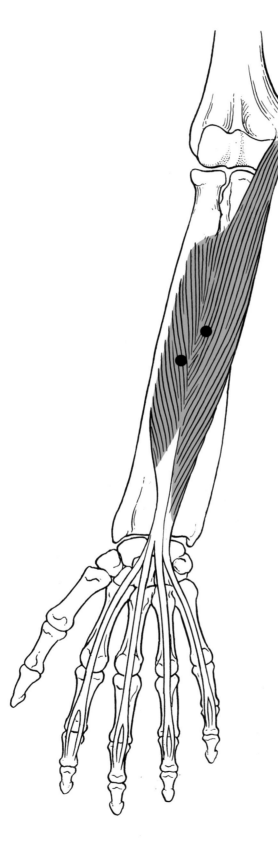

Origin: Humeral head—medial epicondyle of the humerus through the common tendon and the medial margin of the coronoid process of the ulna

Radial head—anterior surface of the shaft of the radius

Insertion: Four tendons divide into two slips each. The slips insert into the sides of the middle phalanges of the four fingers

Action: Flexes the wrist and the middle phalanges of fingers two through five

Innervation: Median nerve (C7, C8, T1)

The tendon of this muscle can be **palpated** on the anterior surface of the wrist between the **flexor carpi ulnaris tendon** and the **palmaris longus tendon** during active flexion. The median nerve and ulnar artery pass beneath the origin of this muscle. At the wrist the tendons of fingers three and four are superficial to the tendons of fingers two and five. Each of the tendons divide at the proximal phalanx to allow the tendon of the flexor digitorum profundus to pass through. The **trigger points** are in the belly of each head of the muscle.

Flexor Digitorum Profundus

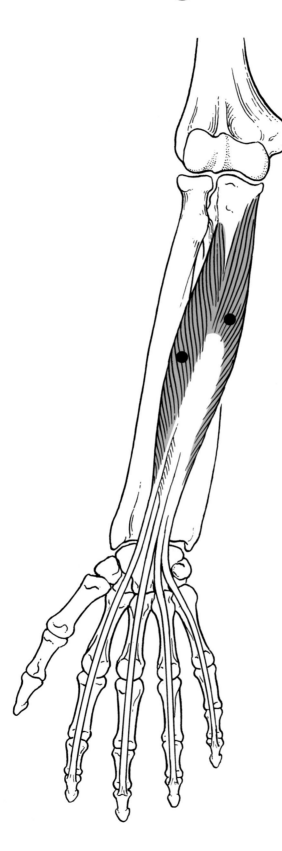

Origin: Medial and anterior surfaces of the proximal three fourths of the ulna and the interosseous membrane

Insertion: By four tendons into the anterior surface of the distal phalanges of digits two through four

Action: Flexes the distal interphalangeal joints of digits two through five and assists in the adduction of the index, ring and little fingers and in flexion at the wrist

Innervation: Ulnar nerve and interosseous branch of the median nerve (C7, C8, T1)

The tendons of this muscle pass through the tendons of the **flexor digitorum superficialis.** Tendons can be **palpated** on the anterior surface of the middle phalanges of the four fingers during active flexion. The **trigger points** are in the belly of the muscle. Its **referred pain pattern** is the wrist into the fingers.

Flexor Pollicis Longus

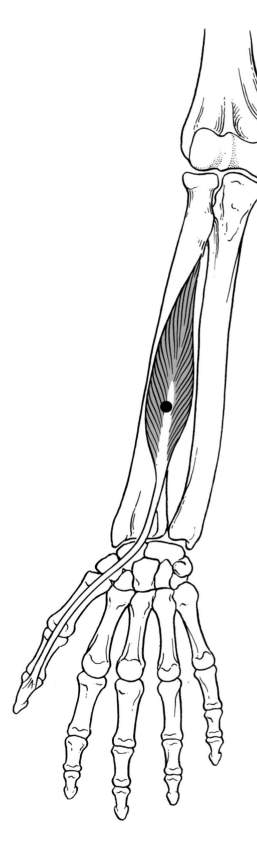

Origin: Middle of anterior shaft of the radius, interosseous membrane, medial epicondyle of humerus and the coronoid process of ulna

Insertion: Palmar surface of the base of the distal phalanx of the thumb

Action: Flexes the thumb

Innervation: Anterior interosseous branch of the median nerve (C8, T1)

The tendon of this muscle can be **palpated** on the anterior surface of the proximal phalanx during active flexion of the thumb. The **trigger point** is in the belly of the muscle. Its **referred pain pattern** is in the wrist and thumb.

Pronator Quadratus

Origin: Medial side of the anterior surface of the distal one fourth of the ulna

Insertion: Lateral side of anterior surface of the distal one fourth of the radius

Action: Pronation of the forearm

Innervation: Anterior interosseous branch of the median nerve (C8, T1)

The **pronator quadratus** is the deepest muscle of the distal forearm. It is the only muscle that arises solely on the ulna and inserts solely on the radius. It is the primary pronator of the forearm.

Extensor Carpi Radialis Longus

Origin: Lower third of lateral supracondylar ridge of humerus

Insertion: Dorsal surface of the base of the second metacarpal bone

Action: Extends the wrist and abducts the hand

Innervation: Radial nerve (C6, C7)

The tendon is **palpated** on the dorsal surface of the wrist at the base of the second metacarpal bone. Its **trigger point** is in its belly. This muscle parallels the **brachioradialis** on the lateral forearm.

Extensor Carpi Radialis Brevis

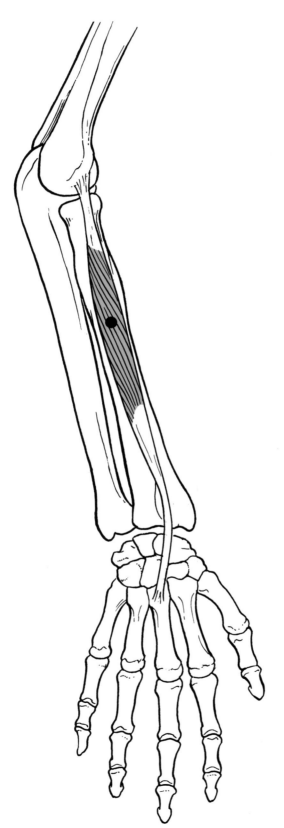

Origin: Lateral epicondyle of the humerus and the radical collateral ligament

Insertion: Dorsal surface of the base of the third metacarpal bone

Action: Extends the wrist and assists in abduction of the hand

Innervation: Radial nerve (C6, C7)

The tendon for the **extensor carpi radialis brevis** can be **palpated** on the dorsal surface of the wrist at the base of the third metacarpal medial to the tendon for the **extensor carpi radialis longus.** Its **trigger point** is in the belly of the muscle. Its **referred pain pattern** is from the lateral epicondyle of the humerus down the posterior portion of the forearm to the hand and the middle finger.

Extensor Pollicis Brevis

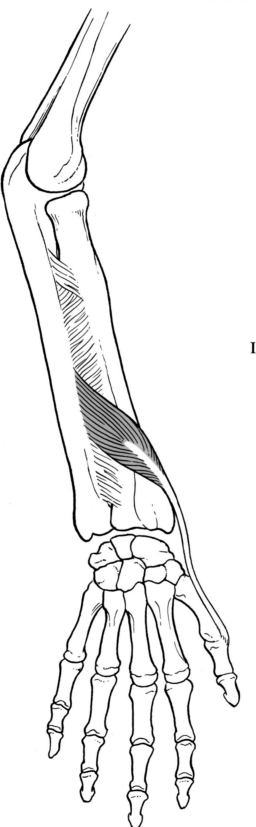

Origin: Dorsal surface of radius and adjacent part of interosseus membrane

Insertion: Extends the proximal phalanx of the thumb

Action: Extends thumb and abducts hand

Innervation: Radial nerve (C6, C7)

Extensor Digitorum

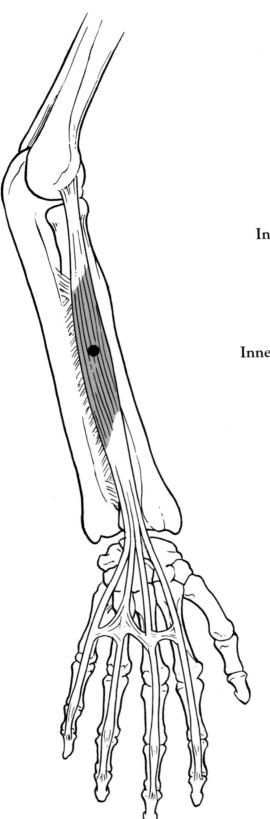

Origin: Common extensor tendon from the lateral epicondyle of humerus

Insertion: By four tendons to the lateral and dorsal surfaces of all the phalanges of digits two through five

Action: Extends the fingers and the wrist

Innervation: Deep branch of radial nerve (C6–C8)

This muscle divides into four prominent tendons on the dorsum of the hand. It can be **palpated** on the middle of the dorsal forearm during forced finger and wrist extension. The tendons can readily be **palpated** on the dorsum of the hands. This muscle tends to hyperextend the metacarpophalangeal joint. Inflammation of this common tendon of origin is a primary cause of **tennis elbow**. The **trigger point** is in the belly of the muscle.

An Illustrated Atlas of the Skeletal Muscles

Extensor Digitis Minimi

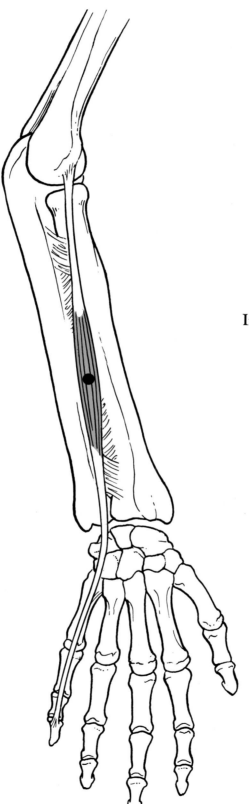

Origin: Common tendon from the lateral epicondyle of the humerus

Insertion: Dorsal surface of base of first phalanx of the fifth digit

Action: Extends the fifth digit

Innervation: Radial nerve (C6–C8)

The tendon of this muscle can be **palpated** on the ulnar side of the tendon of the **extensor digitorum** tendon during resisted extension. The **trigger point** is in the belly of the muscle.

Extensor Carpi Ulnaris

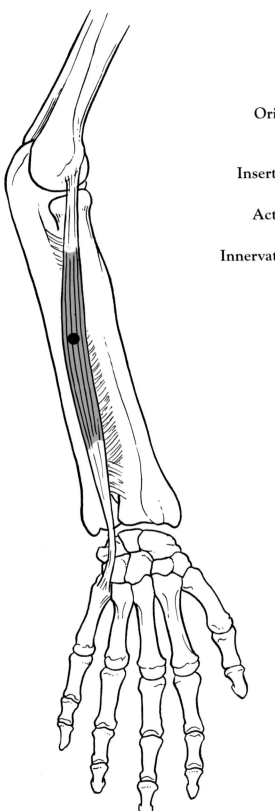

Origin: Common tendon from the lateral epicondyle of the humerus

Insertion: Dorsal surface of base of fifth metacarpal bone

Action: Extends the wrist and assists in adduction of hand

Innervation: Interosseous branch of the radial nerve (C6–C8)

The tendon of the **extensor carpi ulnaris** runs through a groove between the head of the ulna and the styloid process of the ulna. The tendon can be **palpated** on the dorsal surface of the wrist on the ulnar side of the carpal bones. Its **trigger point** is in its belly. Its **referred pain pattern** is from the lateral epicondyle down the posterior surface of the forearm to the little finger.

Supinator

Origin: Lateral epicondyle of humerus, annular and radial collateral ligaments and supinator crest of ulna

Insertion: Lateral surface of the upper one third of the body of the radius

Action: Supinates the forearm

Innervation: Radial nerve (C6)

The **supinator** is a large muscle that wraps around the bones of the forearm. This muscle is covered by the more superficial muscles. Its **trigger point** is near the radius in the antecubital space. Its **referred pain pattern** is from the lateral epicondyle to the dorsal web of the thumb. It mimics tennis elbow. The radial nerve passes between the superficial and deep layers of this muscle.

Abductor Pollicis Longus

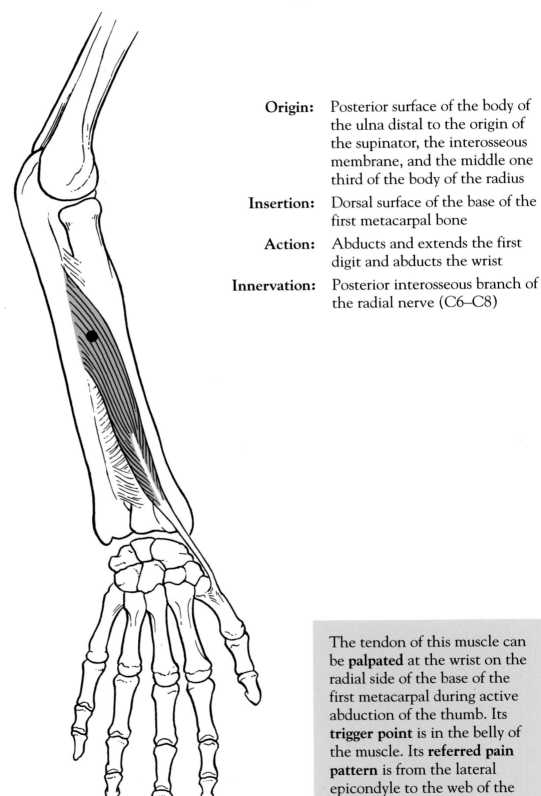

Origin: Posterior surface of the body of the ulna distal to the origin of the supinator, the interosseous membrane, and the middle one third of the body of the radius

Insertion: Dorsal surface of the base of the first metacarpal bone

Action: Abducts and extends the first digit and abducts the wrist

Innervation: Posterior interosseous branch of the radial nerve (C6–C8)

The tendon of this muscle can be **palpated** at the wrist on the radial side of the base of the first metacarpal during active abduction of the thumb. Its **trigger point** is in the belly of the muscle. Its **referred pain pattern** is from the lateral epicondyle to the web of the thumb.

Extensor Pollicis Longus

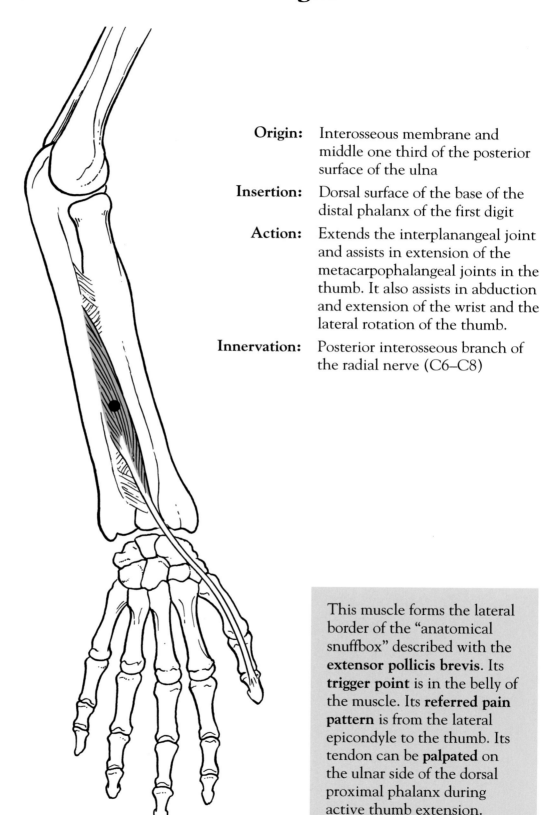

Origin: Interosseous membrane and middle one third of the posterior surface of the ulna

Insertion: Dorsal surface of the base of the distal phalanx of the first digit

Action: Extends the interplanangeal joint and assists in extension of the metacarpophalangeal joints in the thumb. It also assists in abduction and extension of the wrist and the lateral rotation of the thumb.

Innervation: Posterior interosseous branch of the radial nerve (C6–C8)

This muscle forms the lateral border of the "anatomical snuffbox" described with the **extensor pollicis brevis**. Its **trigger point** is in the belly of the muscle. Its **referred pain pattern** is from the lateral epicondyle to the thumb. Its tendon can be **palpated** on the ulnar side of the dorsal proximal phalanx during active thumb extension.

Flexor Pollicis Brevis

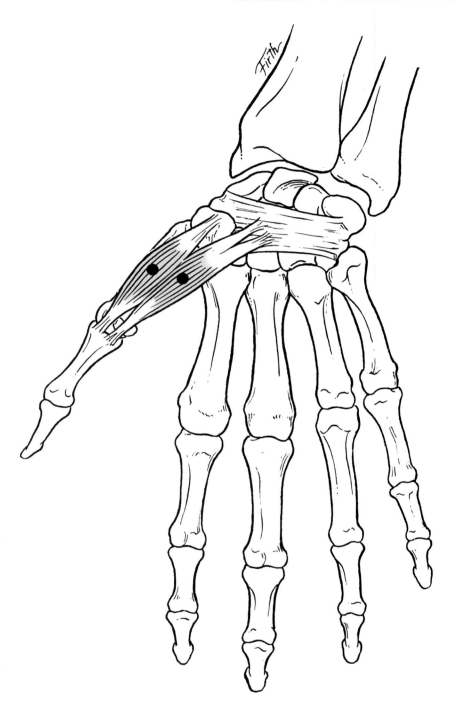

Origin: Flexor retinaculum and trapezium bone

Insertion: Radial side of the base of the proximal phalanx of the thumb

Action: Flexes the proximal phalanx of the thumb

Innervation: Median nerve and deep branch of the ulnar nerve (C8, T1)

The tendon of this muscle can be **palpated** on the anterior surface of the proximal phalanx during active flexion. Its **trigger points** are in the belly of the muscle. The **referred pain pattern** is from the wrist to the thumb.

Extensor Indicis

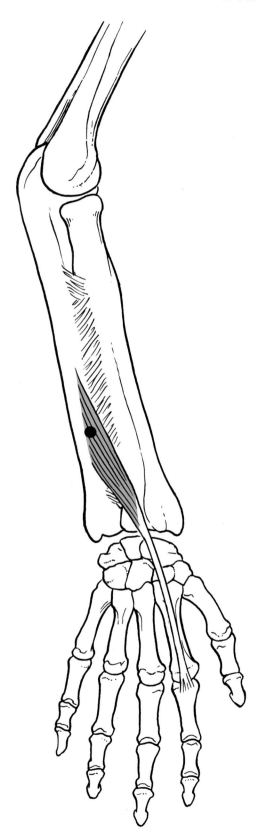

Origin: Interosseous membrane and posterior surface of the ulna

Insertion: Into extensor expansion on dorsal surface of proximal phalanx of the index finger

Action: Extends the index finger at metacarpalphalanx joint

Innervation: Posterior interosseous branch of the radial nerve (C6–C8)

The tendon can be **palpated** in area of base of index finger during resisted finger extension. It is a tiny muscle. The **trigger point** is in the belly of the muscle.

Palmaris Brevis

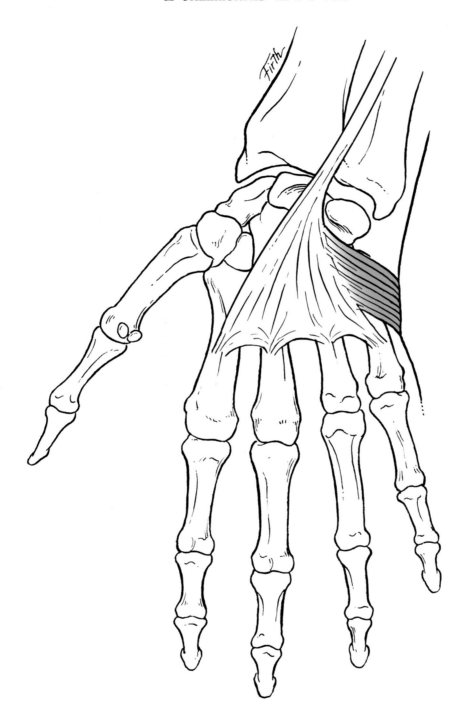

Origin: Palmar aponeurosis

Insertion: Skin of palm of hand

Action: Corrugates skin of palm

Innervation: Ulnar nerve (C8)

This is a small muscle lying in the **fascia** of the **hypothenar eminence**. It is often absent.

Abductor Pollicis Brevis

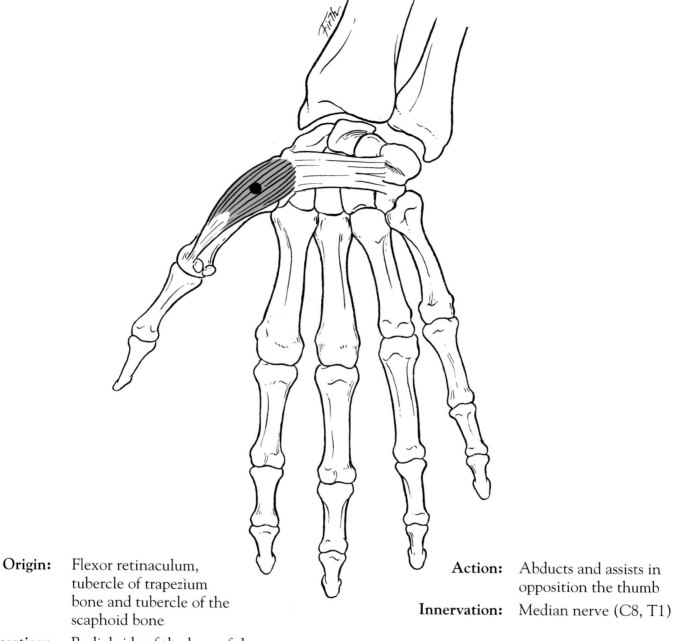

Origin: Flexor retinaculum, tubercle of trapezium bone and tubercle of the scaphoid bone

Insertion: Radial side of the base of the proximal phalanx of the thumb

Action: Abducts and assists in opposition the thumb

Innervation: Median nerve (C8, T1)

This forms the bulk of the muscle on the radial side of palm during active thumb abduction. Its **trigger point** is in the belly of the muscle. Its **referred pain pattern** is the wrist into the thumb. Any lesion that reduces the size of the **carpal tunnel (flexor retinaculum)** may cause compression of the **median nerve**. This causes weakness in the **abductor pollicis brevis** and **opponens pollicis** muscles causing difficulty in performing fine movements with the thumb.

Opponens Pollicis

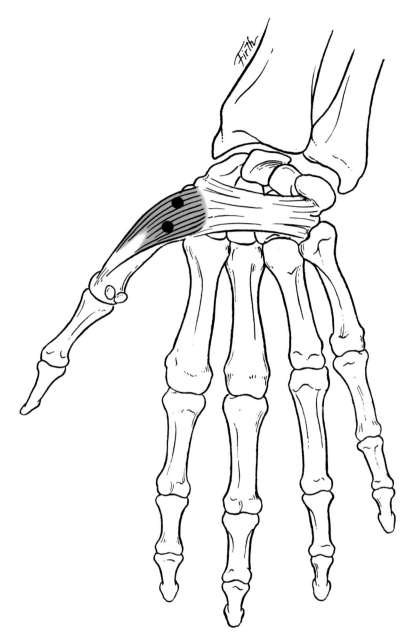

Origin: Ritinaculum and trapezium bone

Insertion: Anterior surface on the radial side of the first metacarpal bone

Action: Rotates thumb into opposition with fingers, acts together with other muscles of the thenar eminence to oppose thumb to fingers

Innervation: Median nerve (C6, C7)

This muscle can be **palpated** along the lateral shaft of the first metacarpal during active thumb opposition. Its **trigger points** are in the belly. Its **referred pain pattern** is from the wrist into the thumb.

Adductor Pollicis

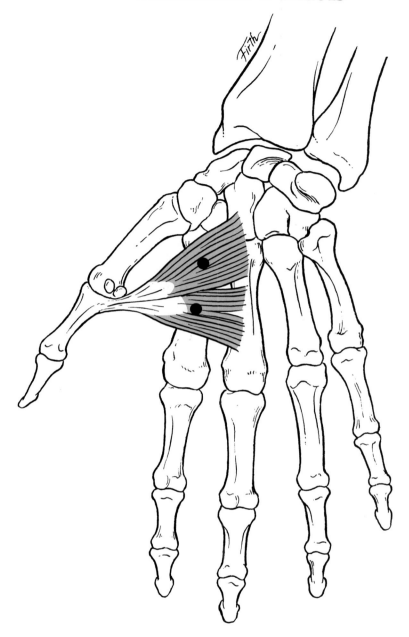

Origin: Oblique head—anterior surfaces of second and third metacarpals, capitate and trapezoid bones

Transverse head—anterior surface of third metacarpal bone

Insertion: Medial side of base of proximal phalanx of thumb

Action: Adducts and flexes thumb

Innervation: Ulnar nerve (C8, T1)

This fan shaped muscle forms the bulk of the thumb web space on the anterior surface. It can be **palpated** deep in the palmar surface of the web space during active adduction of the thumb. Its **trigger points** are in the belly of the muscle. Its **referred pain pattern** is into the thumb.

Abductor Digiti Minimi

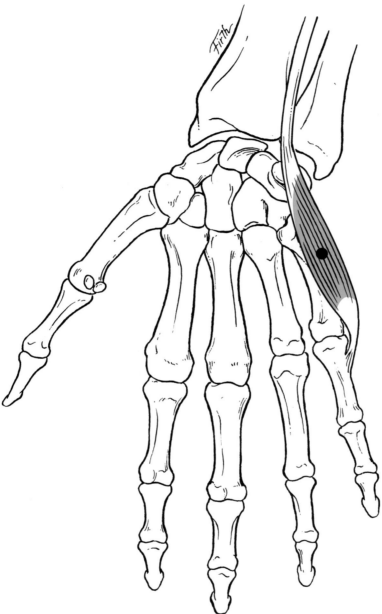

Origin: Pisiform bone and the tendon of the flexor carpi ulnaris

Insertion: Medial side of base of proximal phalanx of the fifth digit

Action: Abducts the fifth finger

Innervation: Ulnar nerve (C8, T1)

This muscle can be **palpated** on the ulnar border of the fifth metacarpal during active abduction of the little finger. The hypothenar eminance (base of the little finger) is less prominent than the thenar eminence (base of the thumb.) Its **trigger point** is in the belly of the muscle. Its **referred pain pattern** is into the little finger.

Flexor Digiti Minimi Brevis

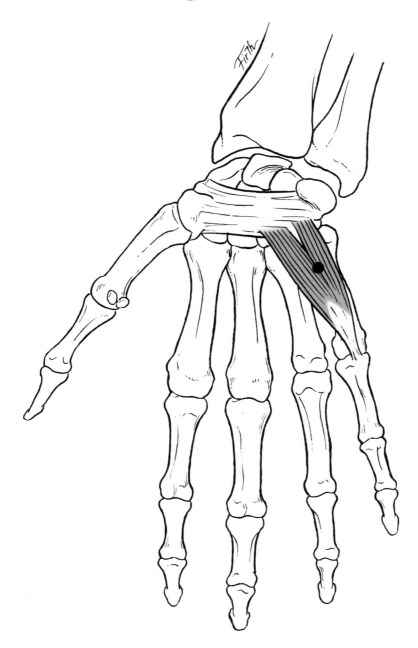

Origin: Flexor retinaculum and hook of the hamate bone.

Insertion: Medial side of base of proximal phalanx of the fifth digit

Action: Flexes fifth finger at metacarpophalangeal joint

Innervation: Ulnar nerve (C8, T1)

This muscle can be **palpated** on the palmar surface of the fifth metacarpal during active flexion of the metacarpophalangeal joint. Its **trigger point** is in its belly and its **referred pain pattern** is into the little finger.

Opponens Digiti Minimi

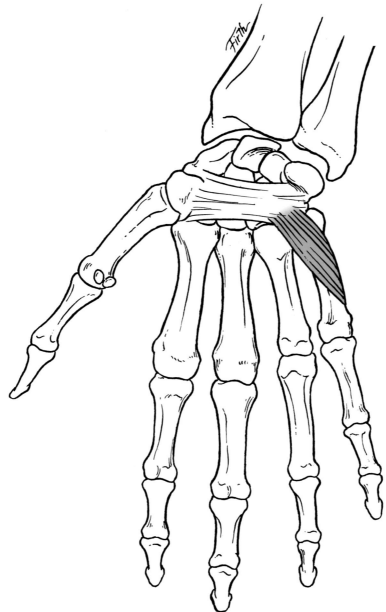

Origin: Anterior surface of flexor retinaculum and hook of hamate

Insertion: Whole length of the medial border of the fifth metacarpal bone

Action: Rotates fifth metacarpal bone into opposition with thumb, draws it forward, and assists in flexing carpometacarpal joint of fifth finger

Innervation: Ulnar nerve (C8, T1)

This muscle helps to cup the palm of the hand. It lies deep to the **abductor digiti minimi** and **flexor digiti minimi** so is difficult to **palpate**.

Lumbricales

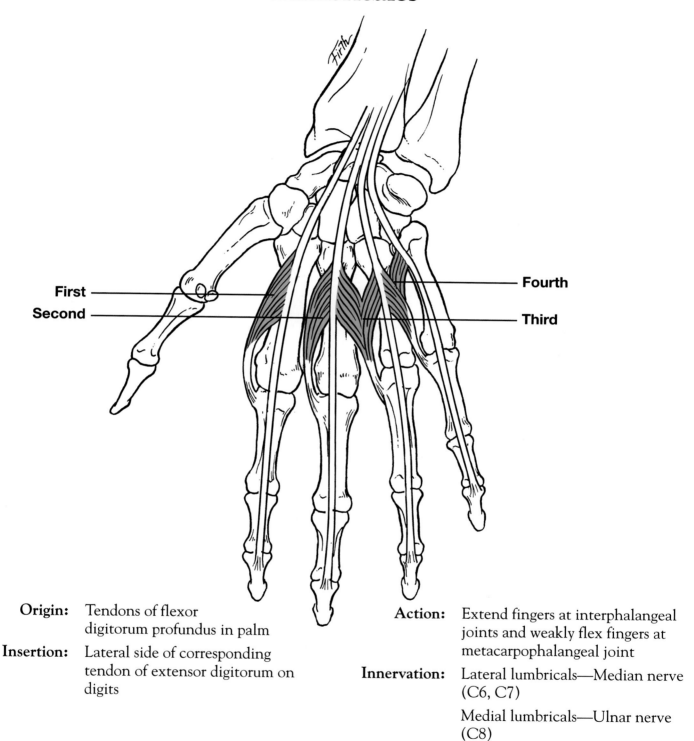

First

Second

Fourth

Third

Origin: Tendons of flexor digitorum profundus in palm

Insertion: Lateral side of corresponding tendon of extensor digitorum on digits

Action: Extend fingers at interphalangeal joints and weakly flex fingers at metacarpophalangeal joint

Innervation: Lateral lumbricals—Median nerve (C6, C7)

Medial lumbricals—Ulnar nerve (C8)

These four muscles assist the **extensor digitorum** in extending the fingers. Simultaneous flexion at the metocarpophalangeal joint and extension at the interphalangeal joints as in holding a cup or pencil is characteristic of these muscles. They cannot be **palpated**.

Dorsal Interossei

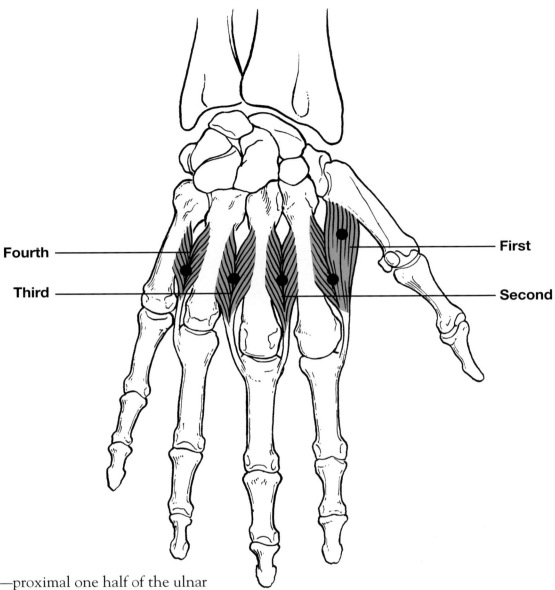

Fourth

Third

First

Second

Origin: First—proximal one half of the ulnar border of the first metacarpal bone and radial border of the second metacarpal bone

Second, third and fourth—adjacent sides of the metacarpal bones in each inter space.

Insertion: First—radial side of the proximal phalanx of the second digit

Second—radial side of the third digit

Third—ulnar side of the third digit

Fourth—ulnar side of the fourth digit

Action: Abducts the index, middle and ring fingers from the midline of the hand.

Innervation: Deep branch of the ulnar nerve (C8, T1)

The **trigger points** are in the belly of the muscles.

Palmar Interossei

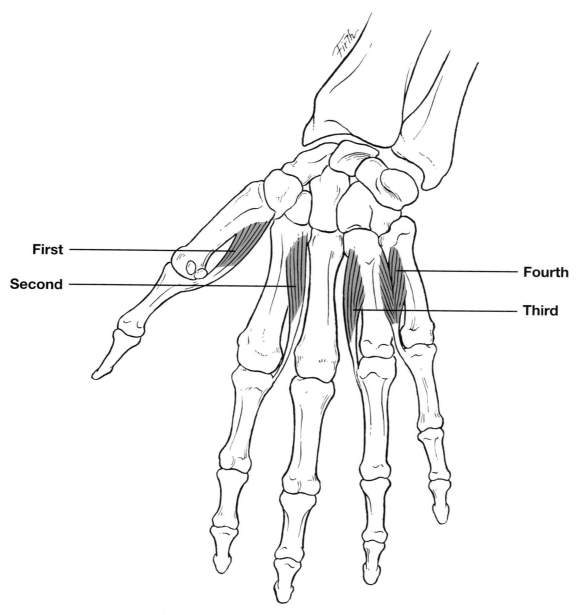

First

Second

Fourth

Third

Origin: First—medial side of base of first metacarpal bone

Second, third and fourth—anterior surfaces of the second, fourth and fifth metacarpal bones

Insertion: First—medial side of base of proximal phalanx of thumb

Second—medial side of base of proximal phalanx of index finger

Third and fourth—lateral side of proximal phalanges of ring finger and fifth finger.

Action: Adducts fingers toward center of third finger at metacarpophalangeal joints and assist in flexion of fingers at the same joints

Innervation: Deep branch of ulnar nerve (C7, T1)

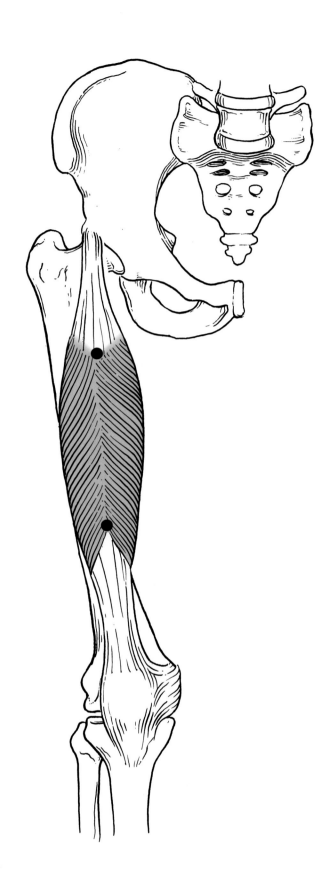

Muscles of the Hip and Thigh

Psoas Major

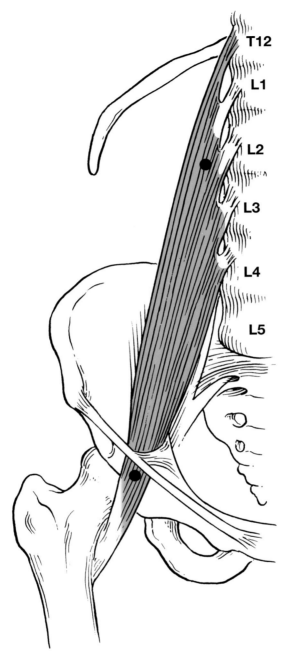

This muscle together with the **iliacus** makes up the **iliopsoas.** It is difficult to **palpate.** Its **trigger points** are near both points of attachment. Its **referred pain pattern** is the entire lumbar area.

Origin: Transverse processes of all lumbar vertebrae, bodies of last thoracic and all lumbar vertebrae, and intervertebral disk of each lumbar vertebrae

Insertion: Lesser trochanter of femur

Action: Flexes thigh at the hip joint and flexes vertebral column

Innervation: Ventral rami of L2, L3, L4

Iliacus

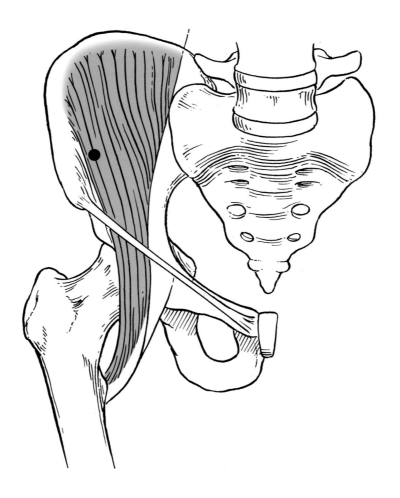

Origin: Upper two thirds of iliac fossa, ala of the sacrum and anterior inferior iliac spine

Insertion: With psoas major, lesser trochanter of femur

Action: Flexes thigh at hip joint

Innervation: Muscular branches of femoral nerve (L2–L4)

This is a large fan-shaped muscle that brings the swinging leg forward in walking or running. Its **trigger point** is near the inner border of the ilium behind the anterior inferior iliac spine. Its **referred pain pattern** is entire lumbar area and front of thigh. It can mimic menstrual pain and appendicitis. The **iliacus** muscle is often considered in conjunction with the psoas as the iliopsoas.

Piriformis

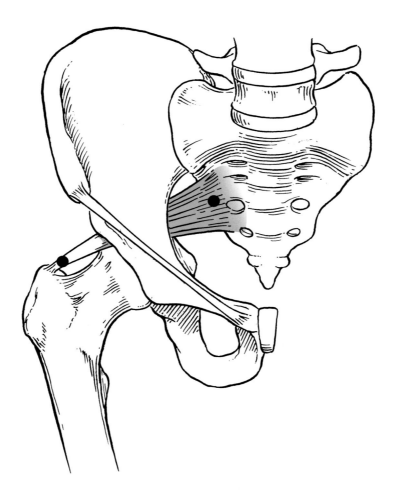

Origin: Pelvic surface of the sacrum between the first through fourth sacral foramina

Insertion: Superior border of the greater trochanter of the femur

Action: Laterally rotates thigh at the hip joint and abducts thigh

Innervation: Anterior rami of S1, S2

This muscle can be **palpated** just posterior to greater trochanter during active lateral rotation of the hip. Its **trigger points** are near the points of attachment. Its **referred pain pattern** is in the sacroiliac region, the entire buttock and down the posterior thigh. Tension in this muscle may cause entrapment of the sciatic nerve which normally passes under the **piriformis** but which in some individuals may pass through the muscle.

Obturator Externus

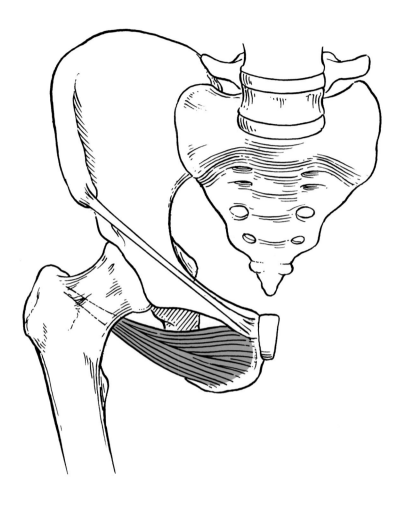

Origin: Anterior surface of superior and inferior rami of pubis and ramus of ischium, and medial side of obturator foramen and external surface of obturator membrane

Insertion: Trochanteric fossa of femur

Action: Laterally rotates thigh at the hip joint

Innervation: Obturator nerve (L3, L4)

This is a flat triangular muscle deep in upper medial aspect of hip under the gluteal muscles.

Obturator Internus

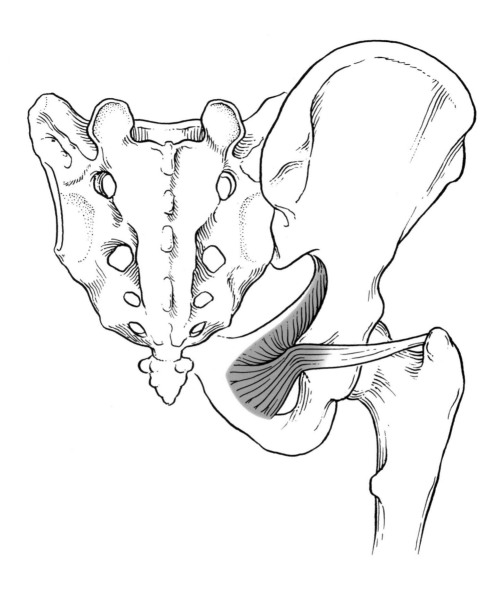

Origin: Pelvic surface of the obturator membrane and the margins of the obturator foramen. Also the internal surface of the pubis and ramus of the ischium

Insertion: Medial surface of the greater trochanter of the femur

Action: Laterally rotates thigh at hip joint

Innervation: L5, S1, S2

This muscle surrounds the obturator foramen in the pelvis. It leaves the pelvis by the lesser sciatic notch and turns sharply forward to insert on greater trochanter. This cannot be **palpated** nor are **trigger points** known.

Gemellus Superior

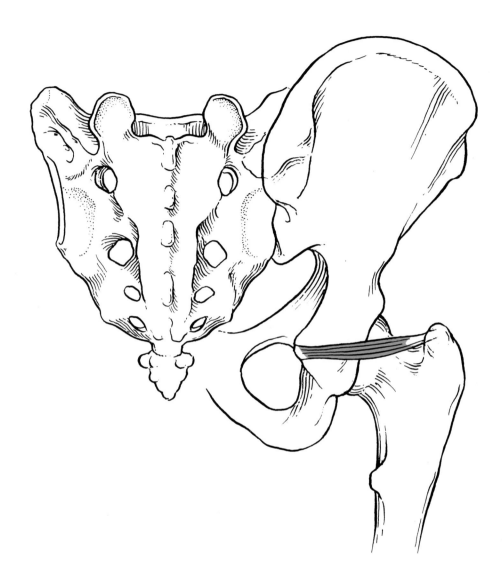

Origin: Dorsal surface of ischial spine

Insertion: With tendon of obturator internus into the upper border of the greater trochanter

Action: Laterally rotates the thigh at the hip joint

Innervation: L5, S1, S2

This is a thin, strap-like muscle.

Gemellus Inferior

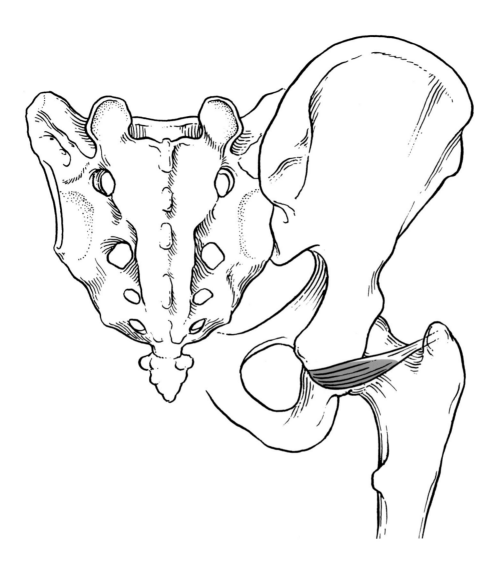

Origin: Upper margin of ischial tuberosity

Insertion: With tendon of obturator internus into upper border of greater trochanter

Action: Laterally rotates thigh at hip joint

Innervation: Nerve from sacral plexus to quadratus femoris (L5, S2)

Quadratus Femoris

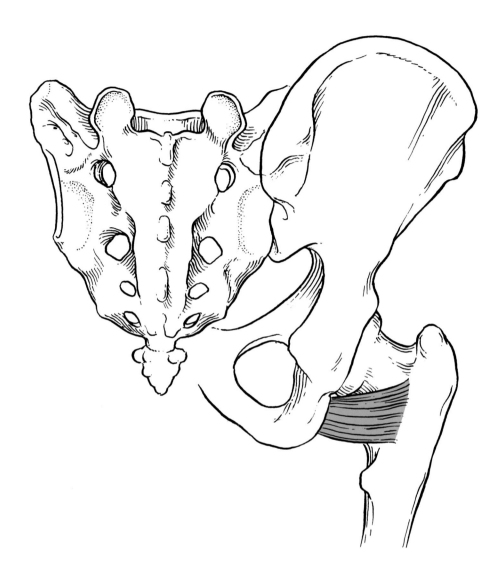

Origin:	Upper part of the lateral border of the ischial tuberosity		**Action:**	Laterally rotates the thigh at the hip joint
Insertion:	Trochanteric crest of femur		**Innervation:**	Branch from sacral plexus (L5, S1)

The **quadratus femoris** can be **palpated** between ischial tuberosity and the greater trochanter during active lateral rotation of the hip. This is the most inferior of the rotators of the hip.

Gluteus Maximus

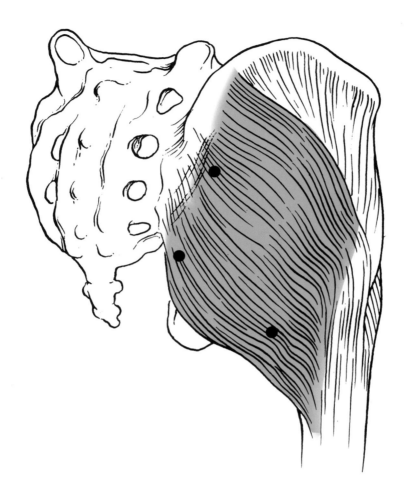

Origin: Posterior gluteal line of ilium, adjacent posterior surface of sacrum and coccyx, sacrotuberous ligament and aponeurosis of erector spinae muscles

Insertion: Iliotibial tract of fascia lata and gluteal tuberosity of femur

Action: Upper part—extends and laterally rotates thigh

Lower part—extends, laterally rotates thigh and assists in raising the trunk from a flexed position. Also assists in adduction of the hip joint.

Innervation: Inferior gluteal nerve (L5, S1, S2)

The **gluteus maximus** muscles are important in maintaining the upright posture. It is active primarily during strenuous activities such as running, jumping, and climbing . It can be **palpated** on the buttock. It has three main **trigger points**: one near the sacrum, one near the ischial tuberosity, and one in the belly of the muscle near the lower fibers. Its **referred pain pattern** is the entire gluteal region. The **gluteus maximus** is an important injection site. The sciatic nerve runs deep through it.

Gluteus Medius

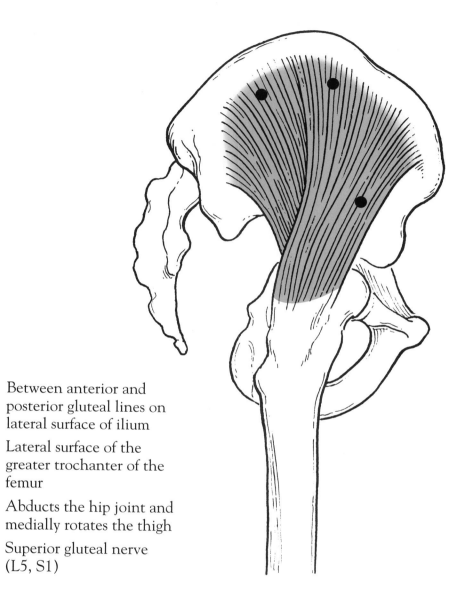

Origin: Between anterior and posterior gluteal lines on lateral surface of ilium

Insertion: Lateral surface of the greater trochanter of the femur

Action: Abducts the hip joint and medially rotates the thigh

Innervation: Superior gluteal nerve (L5, S1)

The gluteal region is a common site for intramuscular injections because the muscles are thick and large. To avoid the sciatic nerve, gluteal nerves, and blood vessels deep in the **gluteus maximus**, injections are applied to the **gluteus medius** in the superolateral part of the buttock where the g. medius is not covered by the g. maximus. The **trigger points** are along the musculotendinous junction at the iliac crest. Its **referred pain pattern** is to the lower back and posterior and lateral areas of the buttock. The muscle can be **palpated** on the lateral aspect of the hip between the iliac crest and the greater trochanter during active abduction.

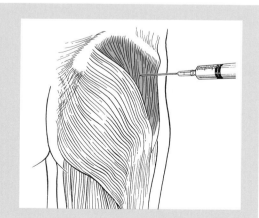

Gluteus Minimus

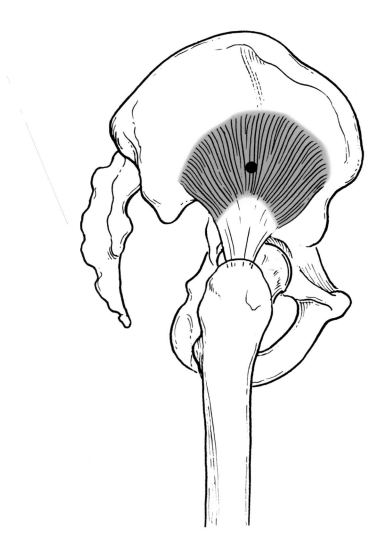

Origin: Outer surface of the ilium between the middle and inferior gluteal lines

Insertion: Anterior border of the greater trochanter

Action: Abducts the femur at the hip joint and medially rotates the thigh

Innervation: Superior gluteal nerve (L4, L5, S1)

This is the smallest and deepest of the gluteal muscles. It can be **palpated** with the **gluteus medius** during active medial hip rotation. The two muscles together prevent the pelvis from dropping toward the opposite side during walking. It also keeps the pelvis level when standing on one foot. Its **trigger point** is in the belly of the muscle. Its **referred pain pattern** is the lower lateral buttock down the lateral aspect of the thigh, lower leg to the ankle.

Tensor Fasciae Latae

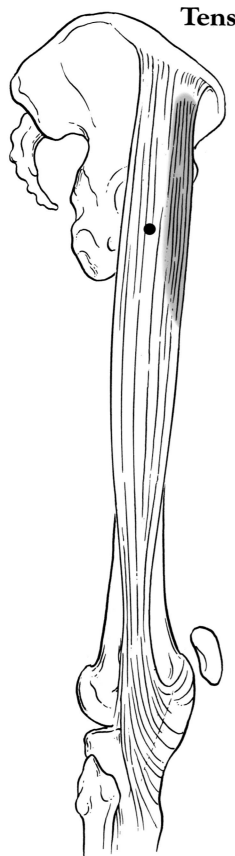

Origin: Anterior aspect of the outer lip of the iliac crest and the anterior superior iliac spine

Insertion: Middle and proximal thirds of the thigh along the iliotibial tract. The iliotibial band inserts on the lateral epicondyle of tibia.

Action: Assists in abduction, medial rotation, and flexion of thigh. Makes the iliotibial tract taut. Stabilizer of the hip.

Innervation: Superior gluteal nerve (L4, L5, S1)

This muscle braces the knee when walking. It can be **palpated** below the superior iliac spine on the anterior pelvis at the level of the greater trochanter during hip abduction. Its **trigger point** is in the belly of the muscle near its insertion. The **referred pain pattern** is localized in the hip and down the lateral side of the leg to the knee.

Sartorius

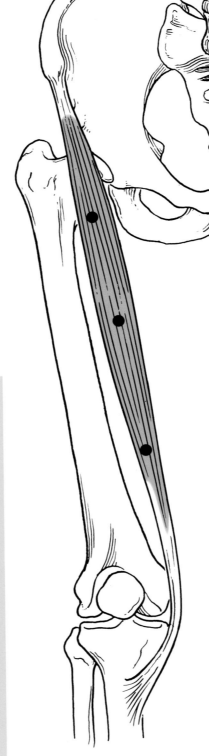

This is a strap-like muscle running obliquely across the anterior surface of the thigh to the knee. It is the longest in the body. It crosses both the hip and knee joint. It can be **palpated** along its length during active flexion, abduction and lateral rotation of the hip. It is sometimes called the **tailor's muscle** and is used in sitting on the floor with thighs spread and lower legs crossed similar to a yoga position. Its **trigger points** are in three or four places in the long belly of the muscle. Its **referred pain pattern** is the entire anterior thigh with concentration at the knee.

Origin: Anterior superior iliac spine and upper half of iliac notch

Insertion: Proximal part of the medial aspect of the tibia

Action: Flexes, laterally rotates, and abducts the hip joint. Also flexes the torso toward the leg, and flexes and assists in medial rotation of the knee.

Innervation: Femoral nerve (L2, L3)

Rectus Femoris

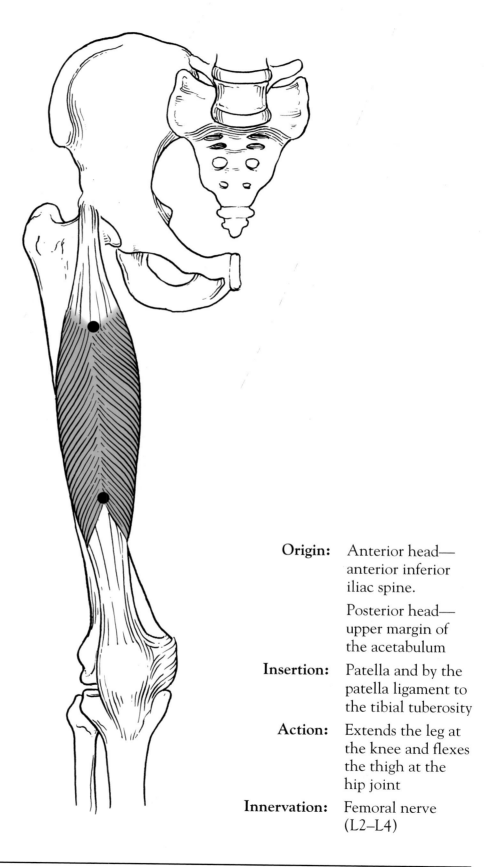

This is one of the four **quadriceps femoris** muscles. It sits on the anterior aspect of the femur and is the only one of the four that crosses both the hip and knee joint. This group is a powerful knee extensor used in running, jumping, climbing and rising from a sitting position. The **rectus femoris** can be **palpated** on the anterior surface of the thigh during active knee extension. It is used when thigh flexion and leg extension are needed such as kicking a soccer ball or football. Its **trigger points** are near its insertion. Its **referred pain pattern** is the entire anterior thigh with a concentration at the knee.

Origin:	Anterior head—anterior inferior iliac spine.
	Posterior head—upper margin of the acetabulum
Insertion:	Patella and by the patella ligament to the tibial tuberosity
Action:	Extends the leg at the knee and flexes the thigh at the hip joint
Innervation:	Femoral nerve (L2–L4)

Vastus Medialis

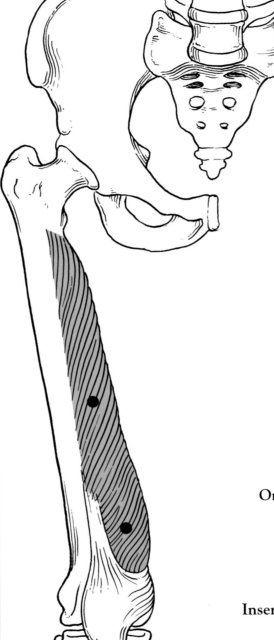

This is the most medial of the muscles of the **quadriceps femoris** group. It forms the medial aspect of the thigh. It can be **palpated** in the anterior-medial surface of the lower third of the thigh during active knee extension. Its **trigger points** are in the belly and just above the insertion. Its **referred pain pattern** is the entire anterior thigh, especially the lower medial aspect with the most concentrated pain in the knee region.

Origin: Lower half of inter-trochanteric line, linea aspera, medial supracondylar line and medial inter-muscular septum

Insertion: Medial border of the patella and then by the patella ligament to the tibial tuberosity

Action: Extends the leg at the knee joint and draws the patella medially

Innervation: Femoral nerve (L2–L4)

Vastus Lateralis

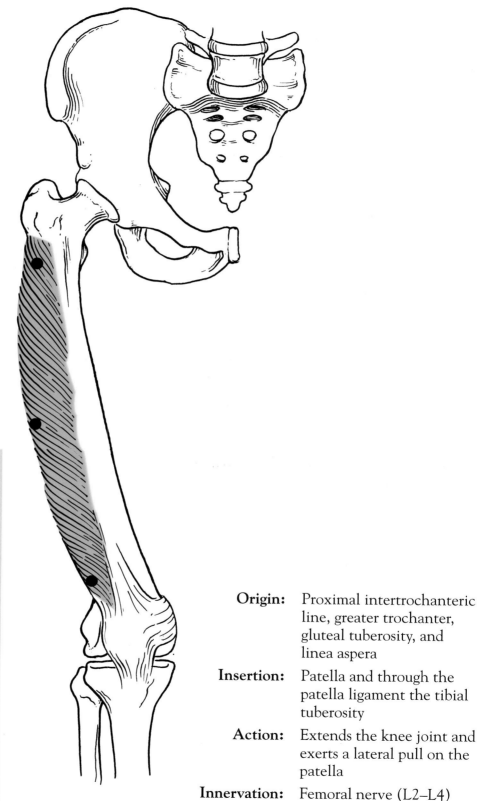

This is the most lateral of the **quadriceps femoris** group. It forms the lateral aspect of the thigh and is a common site for intramuscular injections. It can be **palpated** on the lateral surface of the thigh during active knee extension. Striking the patella ligament causes the characteristic knee jerk reflex test. It has **trigger points** near each attachment and in the belly of the muscle. Its **referred pain pattern** is the anterior thigh especially the lateral surface and again with pain concentrated in the knee.

Origin: Proximal intertrochanteric line, greater trochanter, gluteal tuberosity, and linea aspera

Insertion: Patella and through the patella ligament the tibial tuberosity

Action: Extends the knee joint and exerts a lateral pull on the patella

Innervation: Femoral nerve (L2–L4)

Vastus Intermedius

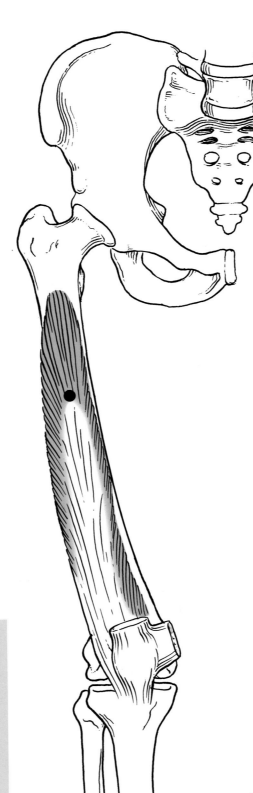

This muscle is the smallest and deepest of the **quadriceps femoris** group. It is covered by the **rectus femoris** and lies between the two **vastus** muscles. It cannot be **palpated** readily. Its **trigger point** is near the insertion. Its **referred pain pattern** is the deep anterior thigh.

Origin: Anterior and lateral surfaces of the proximal two thirds of the body of the femur

Insertion: Deep surface of the tendon of the rectus femoris and vastus muscles. Patella and through the patella ligament to the tibial tuberosity.

Action: Extends the knee at the joint

Innervation: Femoral nerve (L2–L4)

Biceps Femoris

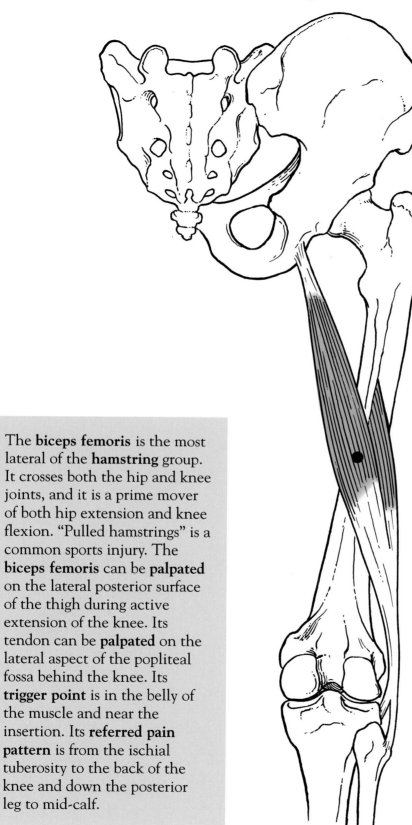

The **biceps femoris** is the most lateral of the **hamstring** group. It crosses both the hip and knee joints, and it is a prime mover of both hip extension and knee flexion. "Pulled hamstrings" is a common sports injury. The **biceps femoris** can be **palpated** on the lateral posterior surface of the thigh during active extension of the knee. Its tendon can be **palpated** on the lateral aspect of the popliteal fossa behind the knee. Its **trigger point** is in the belly of the muscle and near the insertion. Its **referred pain pattern** is from the ischial tuberosity to the back of the knee and down the posterior leg to mid-calf.

Origin: Long head—ischial tuberosity

Short head—lateral lip of linea aspera, proximal two thirds of supracondylar line

Insertion: Common tendon passes downward to insert on head of fibula and lateral condyle of the tibia

Action: Flexes and laterally rotates the knee joint and extends the thigh

Innervation: Long head—tibial division of the sciatic nerve (L5–S2)

Short head—peroneal division of the sciatic nerve (L5, S1, S2)

Semitendinosus

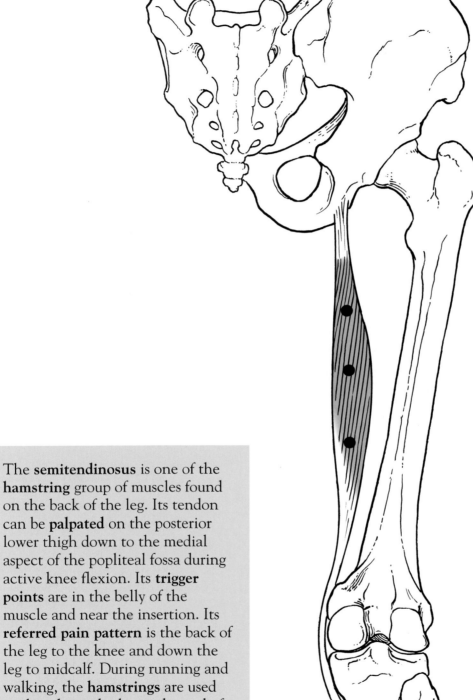

The **semitendinosus** is one of the **hamstring** group of muscles found on the back of the leg. Its tendon can be **palpated** on the posterior lower thigh down to the medial aspect of the popliteal fossa during active knee flexion. Its **trigger points** are in the belly of the muscle and near the insertion. Its **referred pain pattern** is the back of the leg to the knee and down the leg to midcalf. During running and walking, the **hamstrings** are used to slow down the leg at the end of its swing. They are susceptible to being strained by resisting the momentum of the action.

Origin: Ischial tuberosity

Insertion: Upper medial surface of the shaft of the tibia

Action: Flexes and slightly medially rotates leg at knee joint, and extends the thigh at the hip joint

Innervation: Tibial portion of sciatic nerve (L5, S1, S2)

Semimembranosus

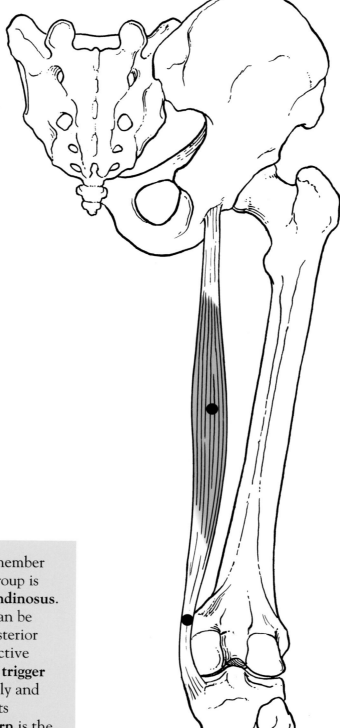

This most medial member of the **hamstring** group is deep to the **semitendinosus**. The muscle belly can be **palpated** on the posterior mid-thigh during active knee extension. Its **trigger points** are in its belly and near its insertion. Its **referred pain pattern** is the back of the thigh, to behind the knee and down the back of the leg to mid-calf.

Origin: Ischial tuberosity

Insertion: Posterior part of the medial condyle of tibia

Action: Flexes and slightly medially rotates leg at knee joint and extends thigh at hip

Innervation: Tibial portion of sciatic nerve (L5, S1, S2)

Gracilis

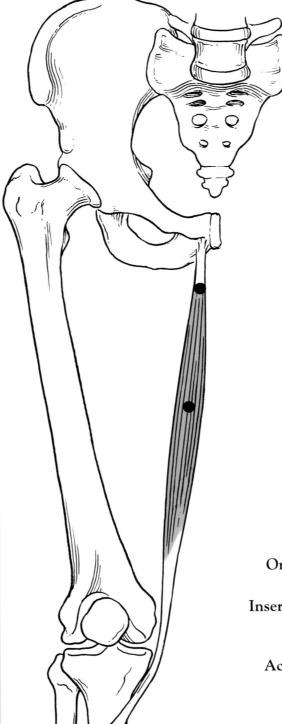

This is a long slender superficial muscle of the medial thigh. It can be **palpated** on the upper medial side of the thigh during active hip adduction. Its tendon can be **palpated** on the medial side of the knee medial to the tendon of the **semitendinosus.** Its **trigger points** are found in the belly of the muscle and near the insertion. Its **referred pain pattern** is deep into the groin, into the medial thigh and downward to the knee and shin.

Origin: Inferior ramus and body of pubis

Insertion: Medial surface of tibia just inferior to its medial condyle

Action: Adducts thigh at hip joint and flexes leg at knee joint. Assists in medial rotation.

Innervation: Obturator nerve (L3, L4)

Pectineus

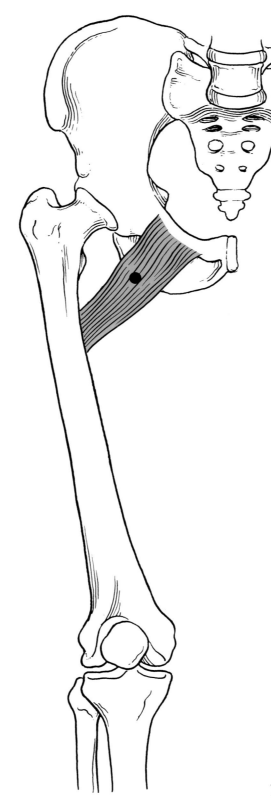

The **pectineus** is the uppermost of the **adductor** group of muscles. There is controversy about whether it medially rotates the thigh. Straining of this muscle causes a "pulled groin" injury. All of the **adductor** group are important in riding or other activities that require the pressing together of the thighs. It can be **palpated** above the **adductor longus** during active thigh adduction. The **trigger point** is in the belly of the muscle. Its **referred pain pattern** is deep in the groin area.

Origin: Pectineal line on superior ramus of pubis

Insertion: From lesser trochanter to linea aspera of femur

Action: Flexes femur at hip and assists in adduction of femur at hip

Innervation: Femoral nerve (L2–L4)

Adductor Brevis

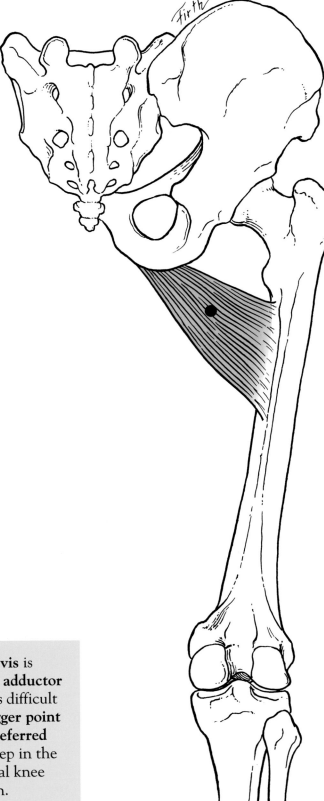

Firth

The **adductor brevis** is found deep to the **adductor longus** and thus is difficult to **palpate**. Its **trigger point** is in its belly. Its **referred pain pattern** is deep in the groin to the medial knee and on to the shin.

Origin: Outer surface of inferior ramus of pubis

Insertion: Upper one third of medial lip of the linea aspera of the femur

Action: Adducts the thigh. Assist in flexion and medial rotation

Innervation: Obturator nerve (L3, L4)

Adductor Longus

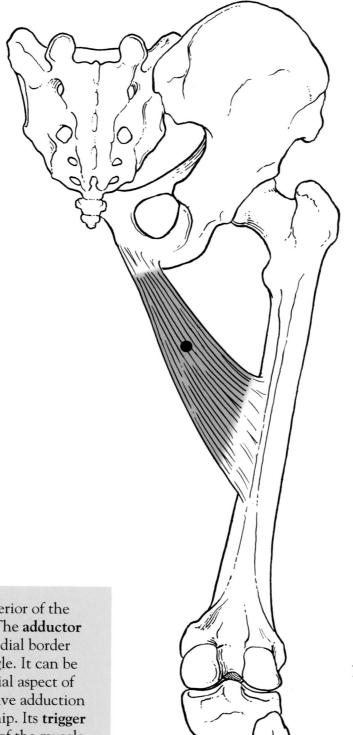

This is the most anterior of the **adductor muscles**. The **adductor longus** forms the medial border of the femoral triangle. It can be **palpated** in the medial aspect of the groin during active adduction of the femur at the hip. Its **trigger point** is in the belly of the muscle. Its **referred pain pattern** is in the groin area downward to the knee and shin.

Origin: Anterior body of pubis

Insertion: Medial one third lip of linea aspera of femur

Action: Adducts and flexes thigh. Assists in medial rotation.

Innervation: Obturator nerve (L2–L4)

Adductor Magnus

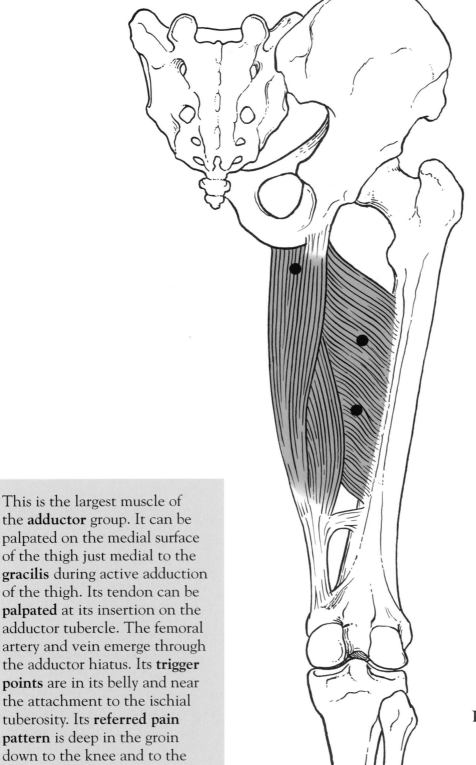

This is the largest muscle of the **adductor** group. It can be palpated on the medial surface of the thigh just medial to the **gracilis** during active adduction of the thigh. Its tendon can be **palpated** at its insertion on the adductor tubercle. The femoral artery and vein emerge through the adductor hiatus. Its **trigger points** are in its belly and near the attachment to the ischial tuberosity. Its **referred pain pattern** is deep in the groin down to the knee and to the shin.

Origin: Inferior ramus of pubis and ramus of ischium and inferior portion of ischial tuberosity

Insertion: Linea aspera and adductor tubercle of femur

Action: Adducts and extends thigh. Assists in medial rotation.

Innervation: Obturator and sciatic nerves (L2–L4)

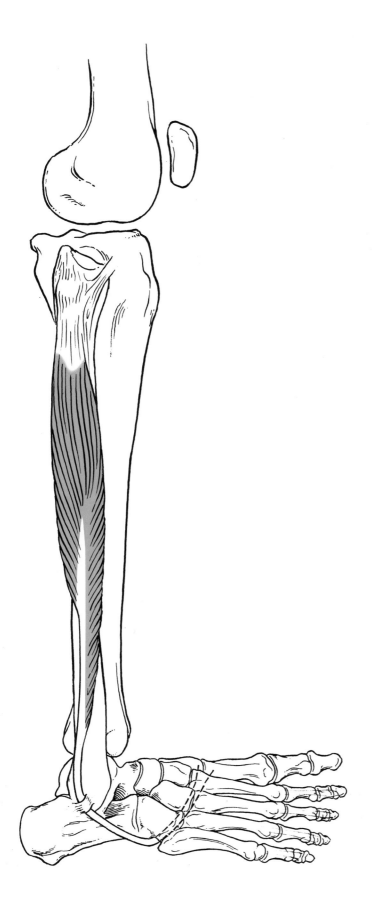

10

Muscles of the Lower Leg and Foot

Soleus

Together the **soleus** and the **gastrocnemius** are referred to as the **triceps surae.** The **soleus** is located beneath the **gastrocnemius**. It can be **palpated** on the lateral side of the lower leg below the belly of the **gastrocnemius** during active plantar flexion. Its tendon is part of the Achilles tendon. Its **trigger points** are near both the origin and insertion. Its **referred pain pattern** is down the entire calf to the heel and sole of the foot into the toes.

Origin: Upper one fourth of posterior surface of the fibula, soleal line, and upper shaft of tibia

Insertion: With the gastrocnemius, via the Achilles tendon to the calcaneous

Action: Plantar flexion of the ankle and stabilizes the leg over the foot

Innervation: Tibial nerve (S1, S2)

POSTERIOR VIEW

An Illustrated Atlas of the Skeletal Muscles

Gastrocnemius

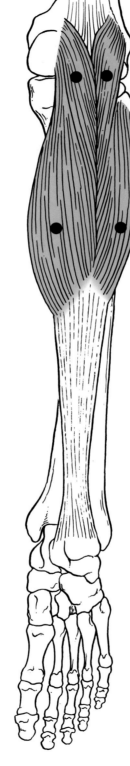

POSTERIOR VIEW

The **gastrocnemius** is the large calf muscle. It can be **palpated** on the posterior calf during active plantar flexion. The Achilles tendon can be **palpated** just above the calcaneous. The **trigger points** are in the belly of the muscle and near the points of origin of each head. Its **referred pain pattern** is down the posterior leg to the heel and sole of the foot into the toes.

Origin: Medial head—upper posterior part of medial condyle of femur

Lateral head—supra-condylar line and lateral condyle of the femur

Insertion: Calcaneous via the Achilles tendon

Action: Plantar flexion of the ankle joint and assists in flexion of the knee joint

Innervation: Tibial nerve (S1, S2)

Plantaris

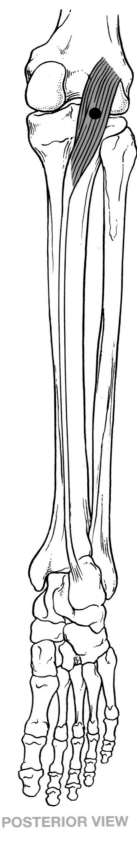

POSTERIOR VIEW

This is a small muscle that is absent in some people. It varies greatly in size. It cannot be **palpated**. Its **trigger point** is in the belly of the muscle behind the knee. Its **referred pain pattern** is behind the knee and into the calf.

Origin: Lateral supracondylar ridge of femur and the oblique popliteal ligament

Insertion: Posterior surface of calcaneous, may be part of Achilles tendon or separate

Action: Plantar flexion of foot

Innervation: Tibial nerve (L4, L5, S1)

Tibialis Anterior

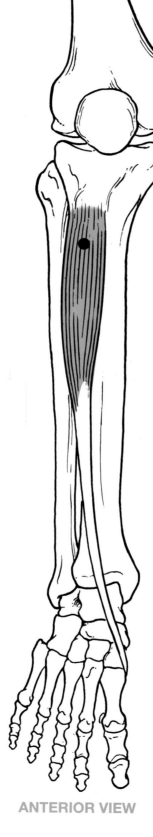

This is a superficial muscle of the shin. It parallels laterally the sharp anterior margin of the tibia. Paralysis of this muscle causes foot drop. Irritation of this muscle during running on hard surfaces often causes "shin splints." The **trigger point** for this muscle would be in its belly. The **referred pain pattern** would be down the leg to the ankle and into the toes. It can be **palpated** on the anterior surface of the tibia during active dorsiflexion. Its tendon can be palpated on the medial side of the anterior surface of the ankle.

ANTERIOR VIEW

Origin: Lateral condyle of and proximal one half of the lateral surface of the tibia and the interosseous membrane

Insertion: Medial plantar surface of the first cuneiform and the base of the first metatarsal bone

Action: Dorsiflexion of the ankle joint and inversion of the foot

Innervation: Deep peroneal nerve (L4, L5, S1)

Extensor Hallucis Longus

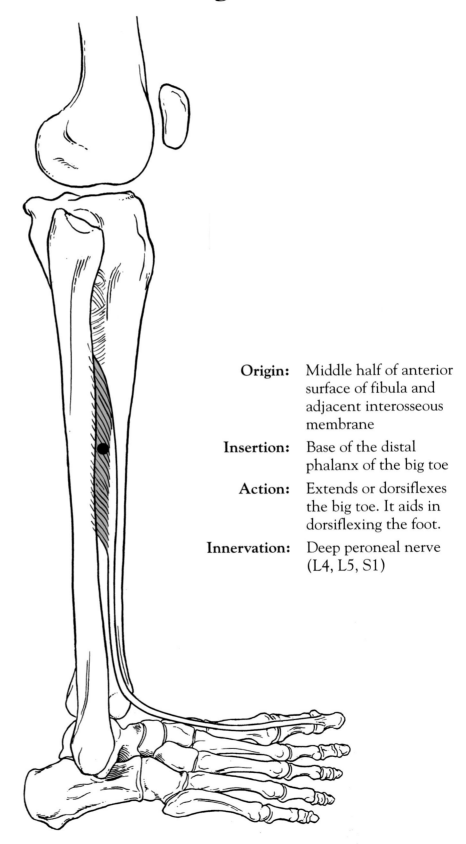

The tendon can be **palpated** lateral to the tibialis anterior tendon on the anterior surface of the ankle and also on the dorsum of the foot near the big toe. Its **trigger point** is in the belly. Its **referred pain pattern** is down the side of the leg into the big toe.

Origin: Middle half of anterior surface of fibula and adjacent interosseous membrane

Insertion: Base of the distal phalanx of the big toe

Action: Extends or dorsiflexes the big toe. It aids in dorsiflexing the foot.

Innervation: Deep peroneal nerve (L4, L5, S1)

Extensor Digitorum Longus

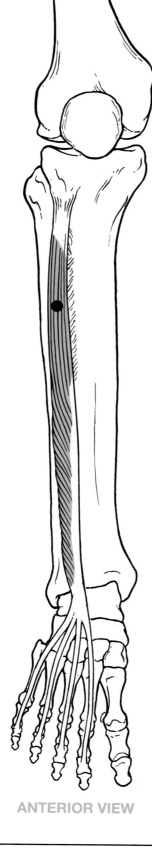

The common tendon can be **palpated** on the anterior surface of the ankle lateral to the **extensor hallucis longus** tendon. The divided tendons can be **palpated** on the dorsum of the foot. The **trigger point** is located near the origin of the muscle, with the **referred pain distribution** mainly to the top of the foot.

ANTERIOR VIEW

Origin: Lateral condyle of the tibia, proximal three fourths of the anterior surface of the fibula, and the interosseous membrane

Insertion: By four tendons to the second through fifth digits. Each tendon divides into a middle slip which inserts in the base of the middle phalanx and two lateral slips which insert in the base of the distal phalanx

Action: Extends the phalanges of the second through fifth digits, assists in dorsiflexion of the ankle and in eversion of the foot

Innervation: Deep peroneal nerve (L4, L5, S1)

Extensor Digitorum Brevis

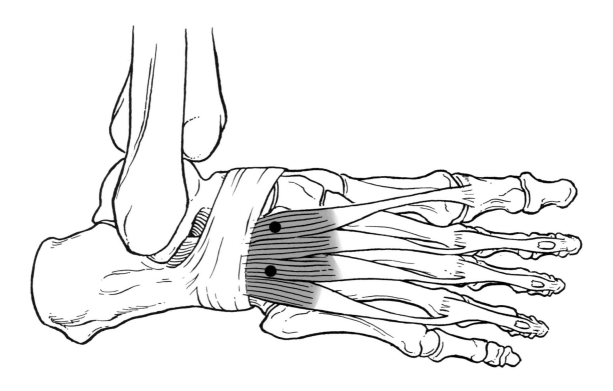

Origin: Anterior and lateral surfaces of the calcaneous, lateral talocalcaneal ligament and inferior extensor retinaculum

Insertion: Dorsal surface of the base of the proximal phalanx of the big toe and the lateral sides of the tendons of the extensor digitorum longus of the second, third and fourth toes

Action: Extends the metatarsalphalangeal joint of the big toe and extends the interphalangeal and metatarsophalangeal joints of the second through fourth toes.

Innervation: Deep peroneal nerve (L5, S1, S2)

The **trigger points** are located toward the origin end of these short toe extensors. The **referred pain pattern** occurs right around these muscles on the outer side of the top of the foot.

Peroneus Longus

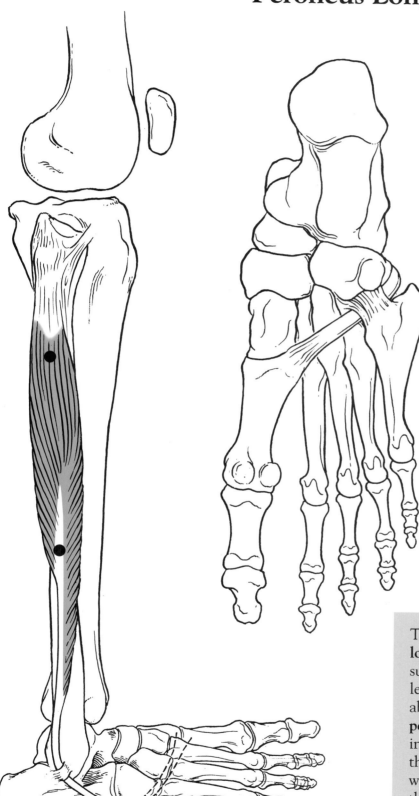

Origin: Upper two thirds of lateral surface of the fibula

Insertion: Lateral side of medial cuneiform and the base of the first metatarsal

Action: Plantar flexion and eversion of the foot

Innervation: Superficial peroneal nerve (L4, L5, S1)

This muscle is also called the **fibularis longus**. It can be **palpated** on the lateral surface of the proximal half of the lower leg, and its tendon can be **palpated** just above the lateral malleolus. Its **trigger points** are located near the origin and insertion. Its **referred pain pattern** is to the lateral malleolus and heel. Together with the **peroneus brevis**, it helps stabilize the lateral ankle and the lateral longitudinal arch of the foot.

Peroneus Brevis

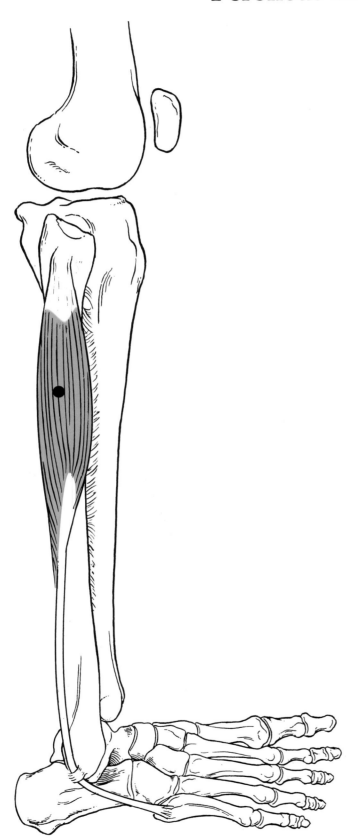

Origin: Lower two thirds of lateral surface of the fibula

Insertion: Lateral side of the base of the fifth metatarsal bone

Action: Plantar flexion and eversion of the foot

Innervation: Superficial peroneal nerve (L4, L5, S1)

This muscle is also called the **fibularis brevis.** The action of the foot "evertors" is helpful when walking or running on uneven surfaces. Its tendon can be **palpated** on the lateral dorsum of the foot near its insertion. Its **trigger point** is in the belly of the muscle. Its **referred pain pattern** is the lateral malleolus.

Peroneus Tertius

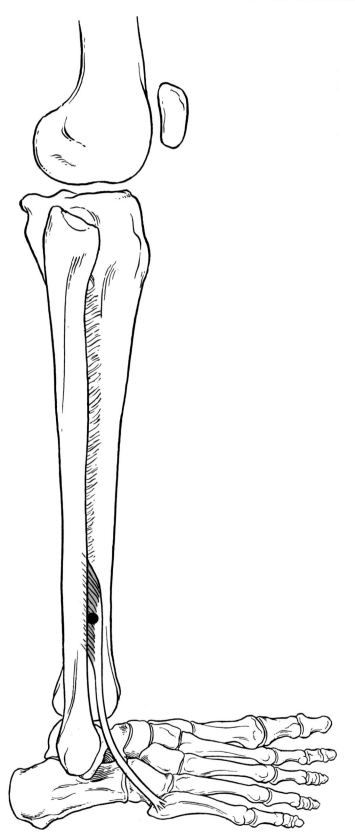

Origin: Lower third of anterior surface of the fibula and the interosseous membrane

Insertion: Dorsal surface of the base of the fifth metatarsal bone

Action: Dorsiflexion and eversion of the foot

Innervation: Deep peroneal nerve (L4, L5, S1)

The tendon of this muscle can be **palpated** lateral to the tendon of the **extensor digitorum longus** on the dorsum of the foot at the base of the fifth metatarsal bone. The **trigger point** is located in the belly of the muscle.

Popliteus

There is some evidence that this muscle stabilizes the knee by preventing lateral rotation of the tibia during medial rotation of the thigh while the foot is planted. It unlocks the knee so that it can be flexed. It is the deepest muscle in the back of the knee. Its **trigger point** is in the belly of the muscle and its **referred pain pattern** is in the back of the knee.

POSTERIOR VIEW

Origin:	Lateral surface of the lateral condyle of the femur
Insertion:	Upper part of the posterior surface of the tibia
Action:	Medially rotates the knee and flexes the leg at the knee
Innervation:	Tibial nerve (L4, L5, S1)

Flexor Hallucis Longus

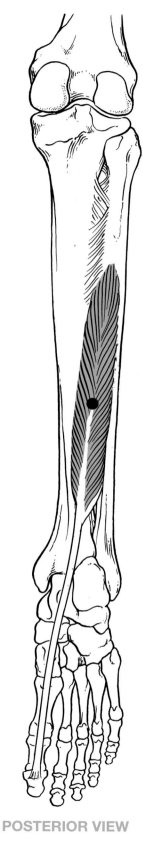

This muscle is important in walking, running and jumping. Its tendon may be **palpated** just medial to the Achilles tendon. Its **trigger point** is in the belly of the muscle. Its **referred pain pattern** is down the posterior leg to the heel and the sole of the foot.

POSTERIOR VIEW

Origin: Lower two thirds of posterior surface of the fibula and the interosseous membrane

Insertion: Plantar aspect of the base of the distal phalanx of the big toe

Action: Flexion of the big toe, plantar flexion of the ankle joint, and inversion of the foot

Innervation: Tibial nerve (L5, S1, S2)

Flexor Digitorum Longus

POSTERIOR VIEW

Its tendon can be **palpated** going around the medial malleolus just posterior to the **tibialis posterior** tendon. Its **trigger point** is in the belly of the muscle. Its **referred pain pattern** is down the posterior leg into the heel.

Origin: Medial part of posterior surface of tibia

Insertion: Plantar surface of the bases of the distal phalanges of the second, third, fourth, and fifth toes

Action: Flexes distal phalanges of lateral four toes, assists in plantar flexion of foot, inverts foot.

Innervation: Tibial nerve (L5, S1)

Tibialis Posterior

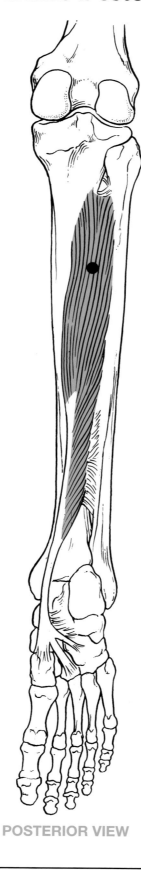

POSTERIOR VIEW

Its tendon can be **palpated** on the medial malleolus during active inversion of the foot. Its **trigger point** is in the belly of the muscle near the knee. Its **referred pain pattern** is in the knee and down the posterior leg.

Origin: Lateral part of posterior surface of tibia, interosseous membrane and proximal half of posterior surface of fibula

Insertion: Tuberosity of navicular bone, cuboid, cuneiforms, second, third and fourth metatarsals and calcaneus

Action: Plantar flexion and inversion of the foot

Innervation: Tibial nerve (L5, S1)

Flexor Digitorum Brevis

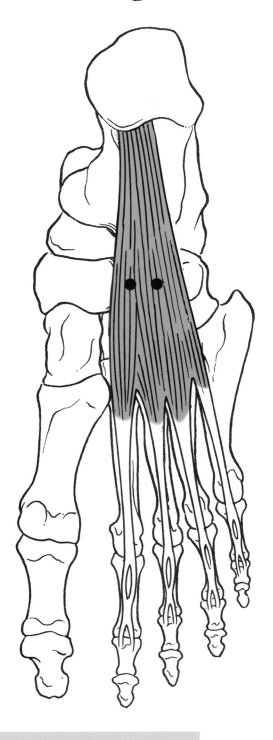

PLANTAR VIEW

This is one of the superficial muscles of the sole of the foot. Together these muscles help support the arches of the foot. The **trigger points** are in the belly of the muscle as it divides into slips onto the 2–5 toes.

Origin: Tuberosity of the calcaneous and plantar apononeurosis

Insertion: Sides of middle phalanges of the second through fifth toes

Action: Flexion of proximal phalanges

Innervation: Medial plantar nerve (L4, L5)

Abductor Hallucis

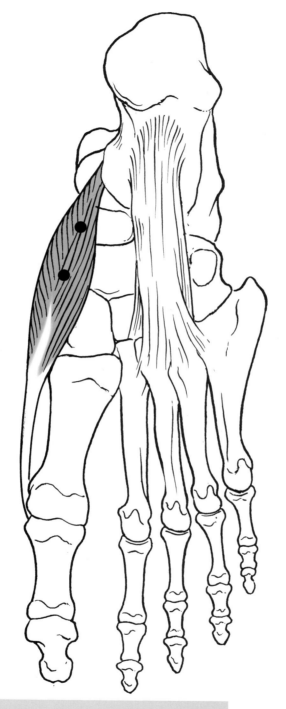

PLANTAR VIEW

The muscles of the sole of the foot can be divided into four
layers from most superficial to deepest. The most superficial
layer includes this muscle and the **flexor digitorum brevis**
and the **abductor digiti minimi**. Its **trigger points** are in the
belly of the muscle.

Origin:	Tuberosity of calcaneous, flexor retinaculum and plantar aponeurosis
Insertion:	Medial side of base of proximal phalanx of big toe
Action:	Abducts and assists in flexion at the metatarsophalangeal joint of the big toe
Innervation:	Medial plantar nerve (L4, L5)

Abductor Digiti Minimi

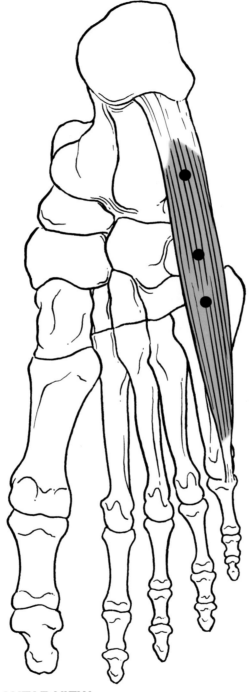

PLANTAR VIEW

This is one of the superficial muscles of the sole of the foot. The **trigger points** are near the origin and in the belly of the muscle.

Origin: Tuberosity of the calcaneous and the plantar aponeurosis

Insertion: Lateral side of proximal phalanx of the fifth toe

Action: Abducts the fifth toe and flexes it at the metatarsophalangeal joint

Innervation: Lateral plantar nerve (S1, S2)

Quadratus Plantae

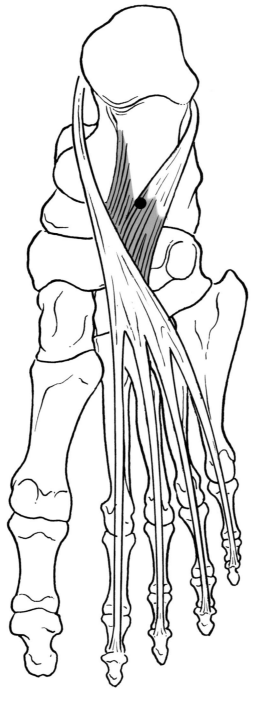

PLANTAR VIEW

This is one of the second layer muscles of the sole of the foot. The **trigger point** is near the attachment to the calcaneous.

Origin: Medial head—medial surface of the calcaneous

Lateral head—lateral border of the inferior surface of the calcaneous

Insertion: Lateral margin of the tendon of the flexor digitorum longus

Action: Flexion of the terminal phalanges of the second through fifth toes

Innervation: Lateral plantar nerve (S1, S2)

Flexor Hallucis Brevis

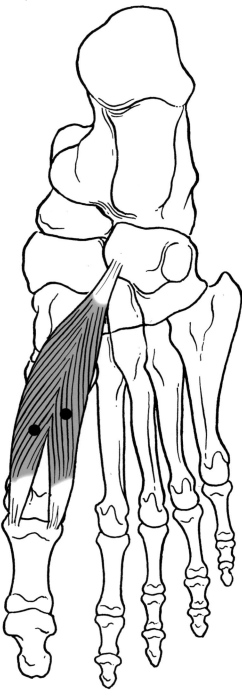

PLANTAR VIEW

This muscle is one of the third layer muscles on the sole of the foot. The **trigger points** are located in the belly of each muscle slip.

Origin: Cuboid and lateral cuneiform bones

Insertion: Medial part—medial side of the base of the proximal phalanx of the big toe

Lateral part—lateral side of the base of the proximal phalanx of the big toe

Action: Flexion of the metatarsophalangeal joint of the big toe

Innervation: Medial plantar nerve (L4, L5, S1)

Lumbricales

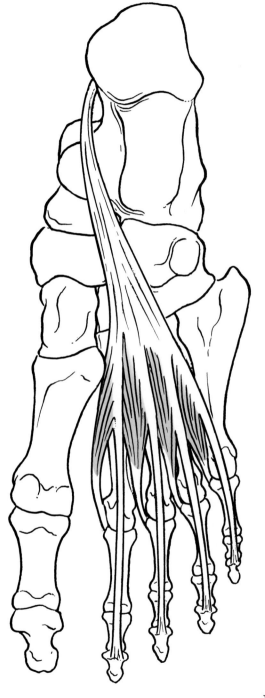

PLANTAR VIEW

This muscle is one of the second layer muscles in the sole of the foot.

Origin: Tendons of the flexor digitorum longus

Insertion: Dorsal surfaces of the proximal phalanges

Action: Flexion of second through fifth toes at the metatarsophalangeal joint

Innervation: First lumbricalis—medial plantar nerve (L4, L5)

Second through fifth lumbricales—lateral plantar nerve (S1, S2)

Flexor Digiti Minimi Brevis

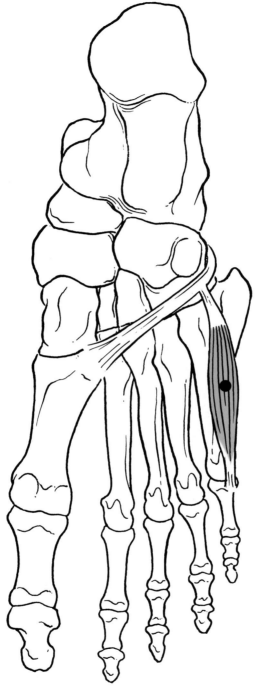

PLANTAR VIEW

Origin: Base of the fifth metatarsal bone and sheath of the peroneus longus tendon

Insertion: Lateral side of base of the proximal phalanx of the fifth toe

Action: Flexion of the proximal phalanx of the fifth toe

Innervation: Lateral plantar nerve (S1, S2)

The **trigger point** is located in the belly of the muscle.

Adductor Hallucis

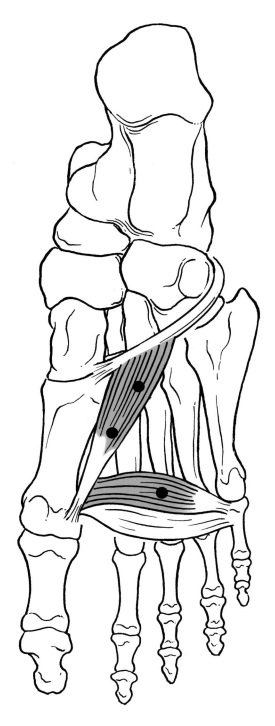

PLANTAR VIEW

Origin: Oblique head—second, third and fourth metatarsal bones and sheath of peroneus longus tendon

Transverse head—ligaments of the third, fourth and fifth toes

Insertion: Lateral side of base of proximal phalanx of the big toe

Action: Adducts the big toe

Innervation: Lateral plantar nerve (S1, S2)

This muscle is one of the third layer muscles in the sole of the foot. It helps to maintain the transverse arch of the foot. The **trigger points** are found in the belly of the transverse head, in the belly and near the insertion of the oblique head.

Plantar Interossei

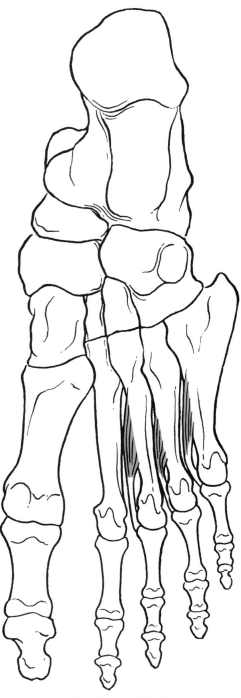

PLANTAR VIEW

Origin: Bases and medial sides of the third, fourth and fifth metatarsal bones

Insertion: Medial sides of bases of proximal phalanges of same toes

Action: Adducts toes and flexion of toes at metatarsalphalangeal joint

Innervation: Lateral plantar nerve (S1, S2)

Dorsal Interossei

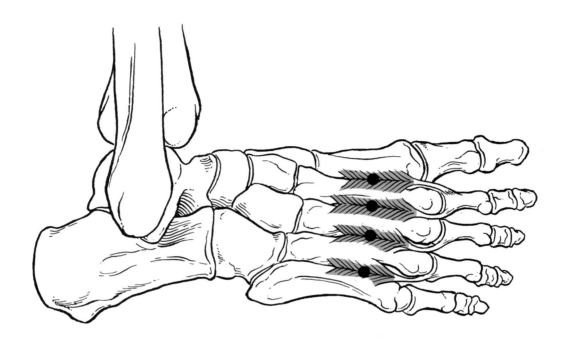

Origin: Adjacent sides of metatarsal bones

Insertion: First—medial side of proximal phalanx of second toe

Second, third and fourth—lateral sides of the bases of proximal phalanges of the second, third and fourth toes

Action: Abduct toes and flexion of proximal phalanges at the metatarsophalangeal joint

Innervation: Lateral plantar nerve (S1, S2)

The **trigger points** are in the belly of each muscle.

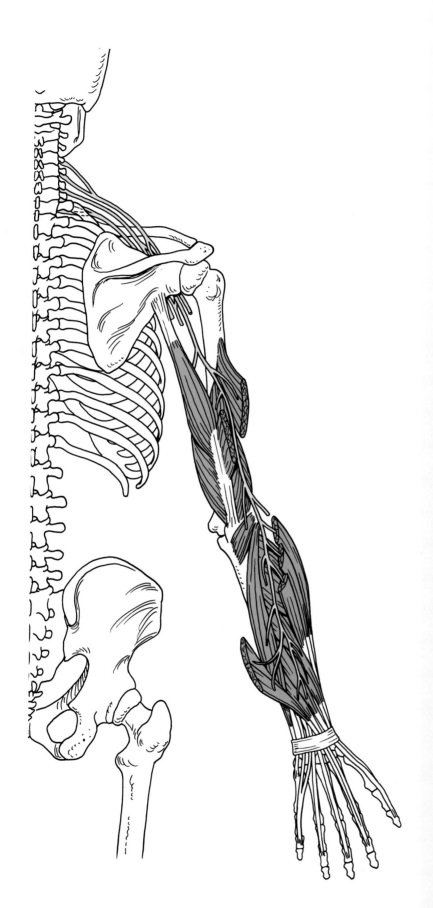

11

Muscle Innervation Pathways

Brachial Plexus

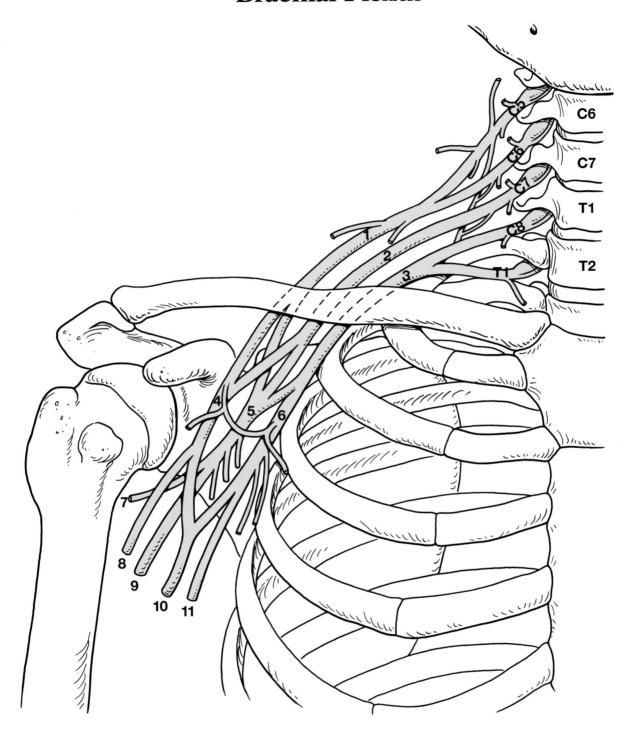

1. Superior trunk
2. Middle trunk
3. Inferior trunk
4. Lateral cord

5. Posterior cord
6. Medial cord
7. Axillary nerve
8. Musculocutaneous nerve

9. Radial nerve
10. Median nerve
11. Ulnar nerve

Axillary (Circumflex) Nerve

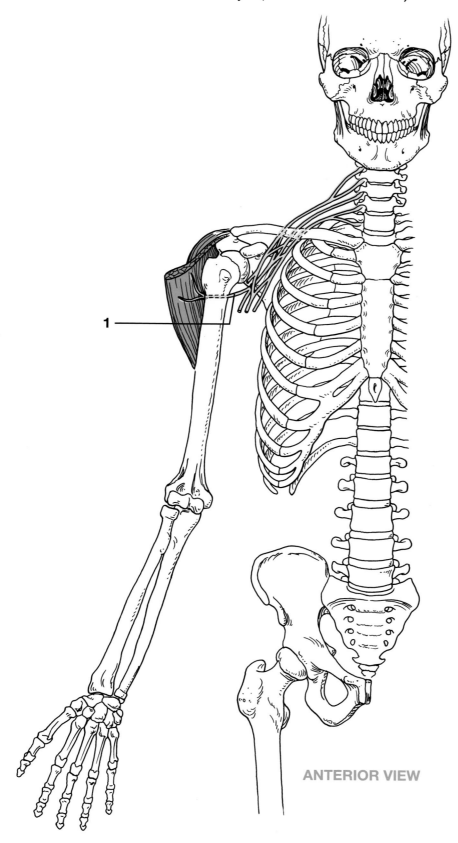

1. Axillary nerve

The **axillary nerve** innervates the **deltoid** and **teres minor** muscles. Those muscles are major movers of the shoulder. The skin in this region is also innervated by this nerve.

ANTERIOR VIEW

Musculocutaneous Nerve

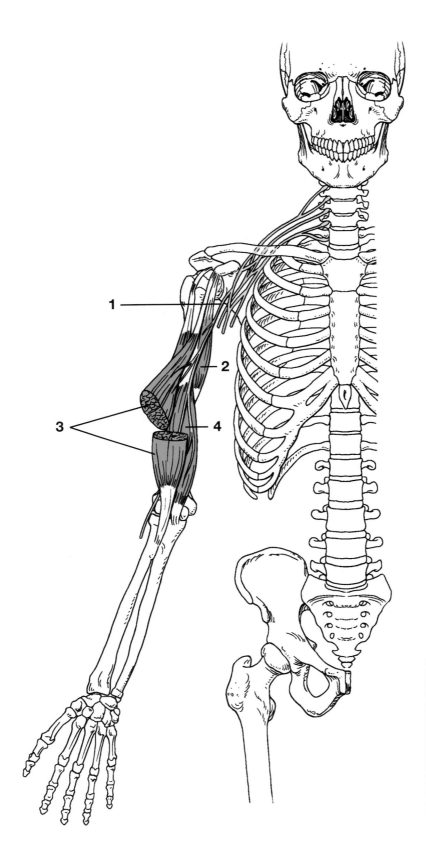

1. Musculocutaneous nerve
2. Coracobrachialis
3. Biceps brachii
4. Brachialis

The **musculocutaneous nerve** innervates the **coracobrachialis, biceps brachii,** and **brachialis** muscles. These muscles are major flexors of the forearm. It also innervates the skin of the anterior surface of the upper arm and forearm.

Radial Nerve

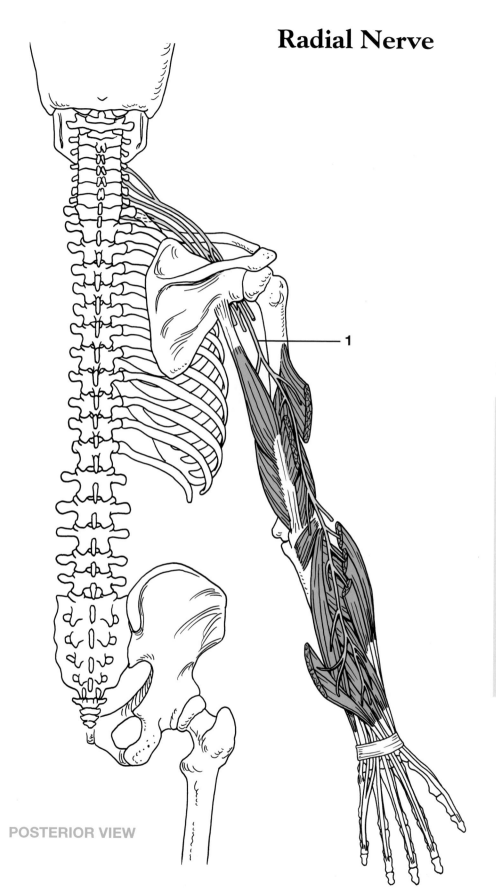

1

POSTERIOR VIEW

1. Radial nerve

The **radial nerve** innervates the **triceps brachii, brachioradialis, anconeus, extensor carpi radialis, supinator, extensor digitorum, extensor digiti minimi, extensor carpi ulnaris, abductor pollicis longus, extensor pollicis brevis,** and **extensor indicis.** These muscles extend the forearm and the hand. It also innervates the skin of the dorsal forearm and hand.

Median Nerve

1. Median nerve

The **median nerve** innervates the **pronator teres, flexor carpi radialis, palmaris longus, flexor digitorum superficialis, flexor pollicis longus, pronator quadratus, abductor pollicis brevis, opponens,** and the **1st and 2nd lumbricals**. These muscles are flexors of the forearm and the hand. The skin of the palmar surface of the hand, thumb, index, and middle fingers are also innervated.

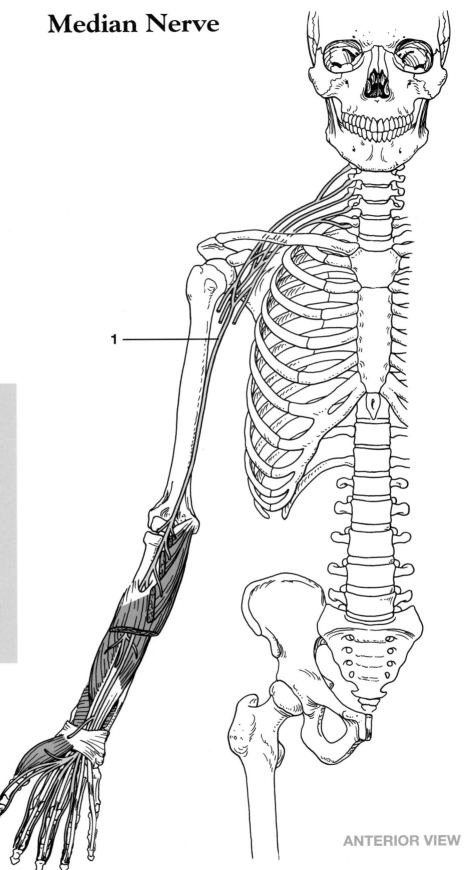

1 ——

ANTERIOR VIEW

An Illustrated Atlas of the Skeletal Muscles

Ulnar Nerve

1. Ulnar nerve

The **ulnar nerve** innervates the **flexor carpi ulnaris, flexor digitorum profundus, adductor pollicis, flexor pollicis brevis, palmar interossei, abductor digiti minimi, opponens digiti minimi, dorsal interossei, and the 3rd and 4th lumbricals.** These muscles flex the hand. It also innervates the skin of the anterior hand and the ring and little fingers.

1

ANTERIOR VIEW

Cutaneous Innervation of the Upper Limb

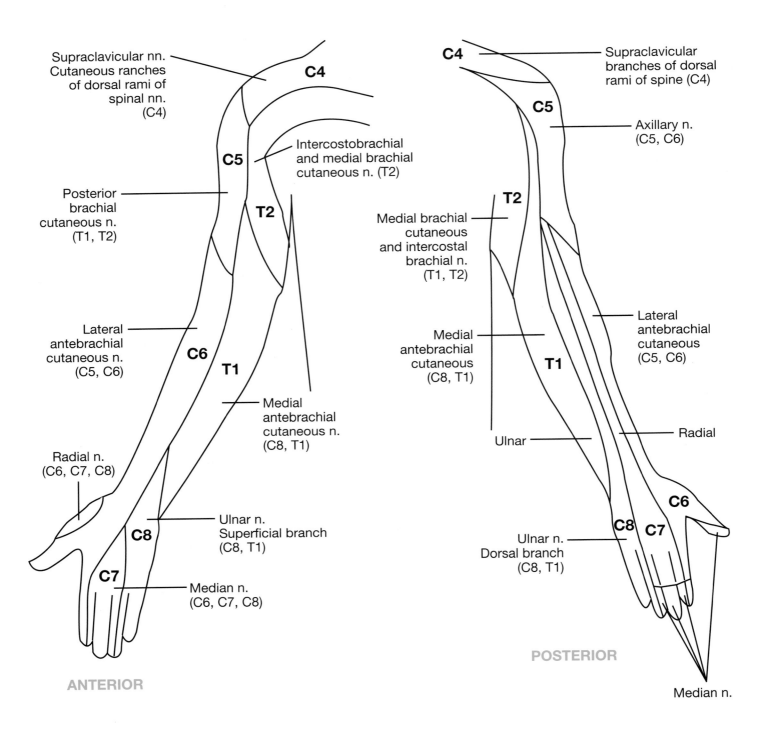

Supraclavicular nn.
Cutaneous ranches
of dorsal rami of
spinal nn.
(C4)

C4

Intercostobrachial
and medial brachial
cutaneous n. (T2)

C5

Posterior
brachial
cutaneous n.
(T1, T2)

T2

Lateral
antebrachial
cutaneous n.
(C5, C6)

C6

T1

Medial
antebrachial
cutaneous n.
(C8, T1)

Radial n.
(C6, C7, C8)

C8

Ulnar n.
Superficial branch
(C8, T1)

C7

Median n.
(C6, C7, C8)

ANTERIOR

C4

Supraclavicular
branches of dorsal
rami of spine (C4)

C5

Axillary n.
(C5, C6)

T2

Medial brachial
cutaneous
and intercostal
brachial n.
(T1, T2)

Lateral
antebrachial
cutaneous
(C5, C6)

Medial
antebrachial
cutaneous
(C8, T1)

T1

Ulnar

Radial

C6

Ulnar n.
Dorsal branch
(C8, T1)

C8 **C7**

POSTERIOR

Median n.

An Illustrated Atlas of the Skeletal Muscles

Lumbosacral Plexus

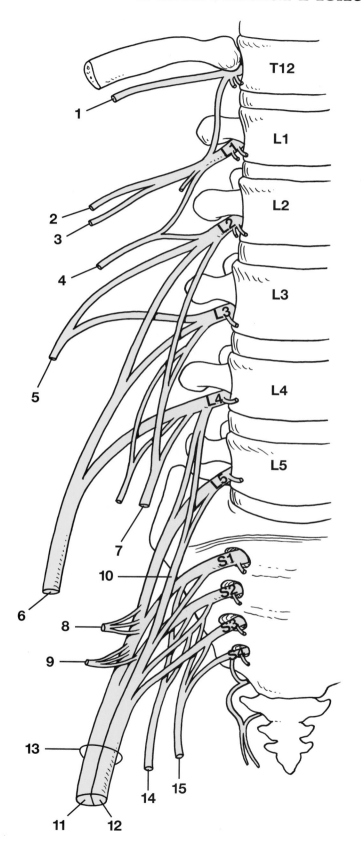

T12

L1

L2

L3

L4

L5

S1

S2

S3

S4

1. 12th thoracic spinal nerve
2. Iliohypogastric nerve
3. Ilioinguinal nerve
4. Genitofemoral nerve
5. Lateral femoral cutaneous nerve
6. Femoral nerve
7. Obturator nerve
8. Superior gluteal nerve
9. Inferior gluteal nerve
10. Lumbosacral trunk
11. Common peroneal nerve
12. Tibial nerve
13. Sciatic nerve
14. Posterior femoral cutaneous nerve
15. Pudendal nerve

Obturator Nerve
Anterior View

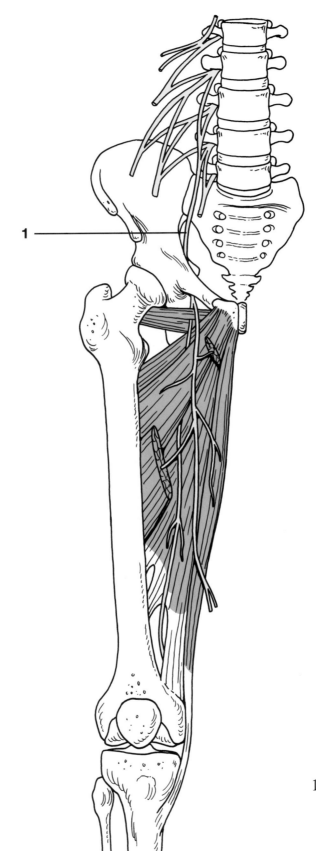

1 ——————————

The **obturator nerve** innervates the **obturator externus, adductor brevis, adductor magnus, adductor longus,** and **gracilis** muscles. These muscles are the major adductors of the upper leg. It also innervates the skin of the medial thigh.

1. Obturator nerve

Femoral Nerve

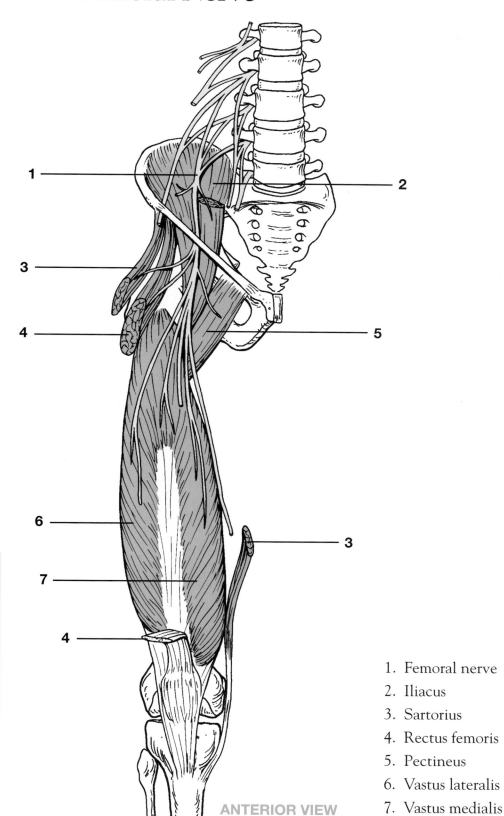

1   2

3

4   5

6   3

7

4

ANTERIOR VIEW

The **femoral nerve** innervates the **iliacus, sartorius, rectus femoris, pectineus, vastus lateralis, vastus intermedius,** and **vastus medialis** muscles. These muscles are hip flexors and extensors of the lower leg. It also innervates the skin of the anterior, lateral, and posterior portions of the thigh.

1. Femoral nerve
2. Iliacus
3. Sartorius
4. Rectus femoris
5. Pectineus
6. Vastus lateralis
7. Vastus medialis

Sciatic Nerve

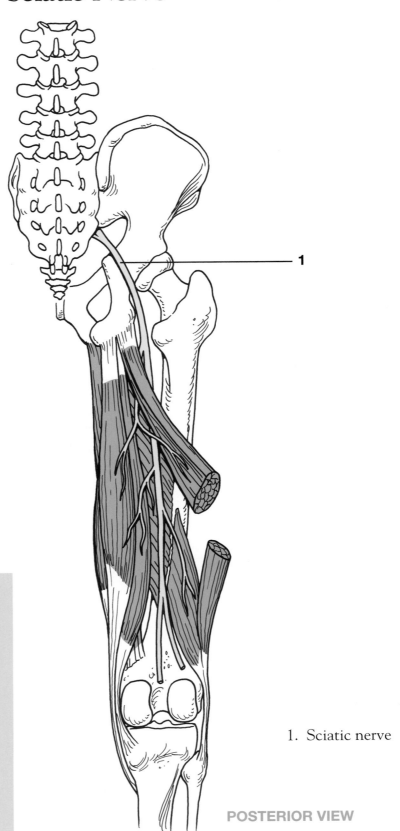

The **sciatic nerve** is composed of two divisions: the **tibial division** and the **peroneal division**. The **tibial division** innervates the **long head of the biceps femoris**, the **semitendinosus**, **semimembranosus** and the posterior portion of the **adductor magnus**. The **peroneal division** innervates the **short head of the biceps femoris**. The **sciatic nerve** divides into the **tibial** and **peroneal nerves** at the popliteal fossa. It also innervates the skin of the leg.

1. Sciatic nerve

POSTERIOR VIEW

An Illustrated Atlas of the Skeletal Muscles

Common Peroneal Nerve

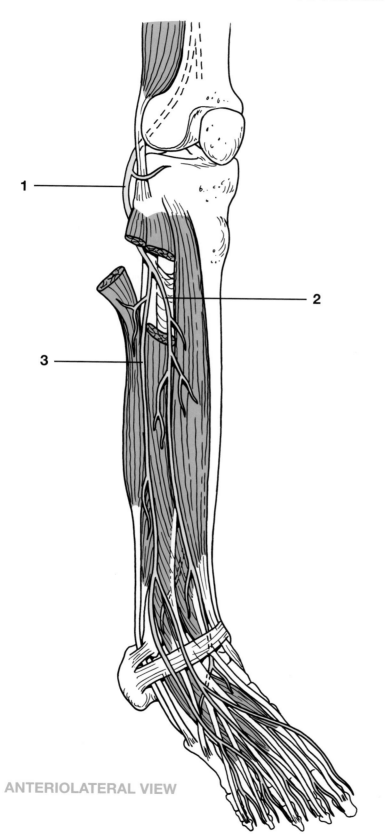

ANTERIOLATERAL VIEW

1. Common peroneal nerve
2. Deep peroneal nerve
3. Superficial peroneal nerve

The **common peroneal nerve** divides into the **deep peroneal nerve** and the **superficial peroneal nerve**. The **superficial peroneal nerve** innervates the **peroneus longus** and **peroneus brevis** muscles. The **deep peroneal nerve** innervates the **tibialis anterior, extensor digitorum longus, extensor hallucis longus, peroneus tertius, extensor hallucis brevis,** and **extensor digitorum brevis** muscles. Together these muscles extend the toes and dorsiflex the foot. It also innervates the skin on the anterior lower leg and the dorsal surface of the foot.

Tibial Nerve

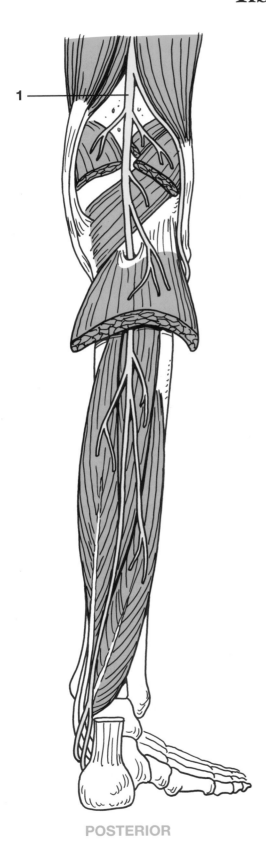

1. Tibial nerve

The **tibial nerve** innervates the muscles of the posterior compartment of the leg including the **plantaris, popliteus, gastrocnemius, soleus, flexor digitorum longus, flexor hallucis longus,** and **tibialis posterior.** One of its branches, the **medial plantar nerve,** innervates the **flexor digitorum brevis, abductor hallucis, flexor hallucis brevis** and **1st lumbrical.** The other branch, the **lateral plantar nerve,** innervates the **adductor hallucis, quadratus plantae, abductor digiti minimi, flexor digiti minimi, plantar interosseous, dorsal interosseous,** and **lumbricals** (lateral 3). It also innervates the skin of the dorsal surface of the lower leg and the plantar surface of the foot.

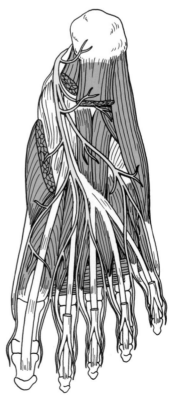

POSTERIOR

PLANTAR

An Illustrated Atlas of the Skeletal Muscles